Overflow Metabolism

Overflow Metabolism
From Yeast to Marathon Runners

Alexei Vazquez
Cancer Research UK Beatson Institute, Glasgow, United Kingdom

ACADEMIC PRESS

An imprint of Elsevier

Academic Press is an imprint of Elsevier
125 London Wall, London EC2Y 5AS, United Kingdom
525 B Street, Suite 1800, San Diego, CA 92101-4495, United States
50 Hampshire Street, 5th Floor, Cambridge, MA 02139, United States
The Boulevard, Langford Lane, Kidlington, Oxford OX5 1GB, United Kingdom

British Library Cataloguing-in-Publication Data
A catalogue record for this book is available from the British Library

Library of Congress Cataloging-in-Publication Data
A catalog record for this book is available from the Library of Congress

ISBN: 978-0-12-812208-2

For Information on all Academic Press publications
visit our website at https://www.elsevier.com/books-and-journals

Working together
to grow libraries in
developing countries

www.elsevier.com • www.bookaid.org

Publisher: Mica Haley
Acquisitions Editor: Linda Versteeg-buschman
Editorial Project Manager: Tracy I. Tufaga
Production Project Manager: Poulouse Joseph
Designer: Matthew Limbert

Typeset by MPS Limited, Chennai, India

aerobic
growth
cancer
metabolism acetate
running
mitochondria efficiency
crowding
overflow
respiration cell optimal
wine ethanol Warburg lactate fermentation
yeast Crabtree speed packing Marathon
beer Pasteur

To my family

CONTENTS

PREFACE

"Nothing in biology makes sense except in the light of evolution"
Theodosius Dobzhansky

You may have heard of overflow metabolism under different names: The Crabtree effect in baker's/brewer's/winemaking yeasts, the lactate switch in sport physiology or the Warburg effect in cancer. You may have also heard about PET scans, a cancer test based upon overflow metabolism. You may have felt it yourself following intense physical activity. That tingling feeling you experience after a sprint is overflow metabolism at work. Could these diverse phenomena be rooted on a common underlying principle?

A universal theory of overflow metabolism has to fit in quite diverse organisms and physiological contexts, from yeasts to Marathon runners. Warburg and Crabtree noticed overflow metabolism when cells increased their metabolism due to cancerous growth or viral infection. Lactate accumulation in blood manifests during intense physical activity. The key point is that metabolism has been pushed to its limits. That leads me to the hypothesis that overflow metabolism and maximum metabolic rates are rooted on the same basic principles. As we shall see, it is all about packing mitochondria in a cell box, or molecular crowding in the language of biophysics.

This book represents an effort to explain overflow metabolism using simple arguments. All conclusions are derived from simple calculations on the back of an envelope. The first chapter is a historical introduction to overflow metabolism and it is accessible to any reader. No previous knowledge of biochemistry is required. Chapter 2, Biochemical Horsepower, introduces some basic concepts about the metabolic capacity of mitochondria and metabolic pathways. While it contains a few formulas, it is also accessible to a general reader. Chapter 3, How Fast Can We Run?, and Chapter 4, How Fast Can We Grow?, focus on overflow metabolism in the context of muscle metabolism and cell growth. These chapters contain some mathematical derivations that, while simple, require some acquaintance with the manipulation of mathematical formulas. In Chapter 5, Overflow Metabolism in Human Disease, problems of fermentation in human diseases are addressed from the point of view of whole body metabolism. Finally, the main conclusions and open questions are highlighted in Chapter 6, Outlook.

ACKNOWLEDGEMENT

I take the opportunity to thank Marcio Argollo de Menezes for the discussions igniting the research on molecular crowding and overflow metabolism. Many thanks to Zoltan N. Oltvai for the several discussions and collaborations that helped shaping these concepts. Finally, I would like to thank Matthias Pietzke, Elke K Markert and Jorge Fernandez de Cossio Diaz for their comments on the first draft of this manuscript.

CHAPTER 1

A Historical View of Overflow Metabolism

En résumé, la fermentation est un phénomène très-général. C'est la vie sans air, c'est la vie sans gaz oxygène libre

Louis Pasteur

1.1 CATABOLISM, ANABOLISM AND OVERFLOW

Metabolism is the biological process where cells, tissues or whole organisms utilize organic molecules for survival, proliferation and to transform the environment. Metabolism is conceptually divided into catabolism and anabolism. *Catabolism* involves the breakdown of organic molecules to generate energy and precursor molecules for anabolism. *Anabolism* involves the synthesis of cell components from precursor molecules and energy. When catabolism is fully efficient, organic molecules are fully incorporated into the cell biomass or they are broken down to carbon dioxide to obtain the maximum energy yield. Under certain conditions, the organic molecules are only partially broken, and the products of incomplete catabolism are released to the environment. This phenomenon of incomplete catabolism, characterized by the excretion of metabolic products that could be otherwise incorporated in the cell biomass or catabolized to carbon dioxide, is called *overflow metabolism.*

Glycolysis, the metabolic pathway comprising the first steps in the catabolism of sugars, is a classical example of the interplay between catabolism, anabolism and overflow metabolism (Fig. 1.1). Sugars are first broken down to glucose and glucose is then catabolized to two molecules of pyruvate. In its anabolic role, pyruvate is used to build larger molecules. Alternatively, pyruvate can be catabolized to carbon dioxide by the metabolic pathway of *oxidative phosphorylation*. The energy released by the catabolism of sugars is used to produce adenine triphosphate (ATP), the energy currency of cells. Glycolysis has a yield of 2 ATP molecules per molecule of glucose. Glycolysis combined with the oxidative phosphorylation of pyruvate has a much higher yield,

Overflow Metabolism. DOI: https://doi.org/10.1016/B978-0-12-812208-2.00001-9

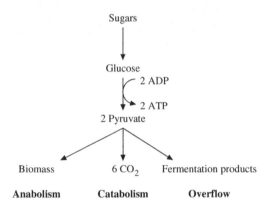

Figure 1.1 Anabolism, catabolism and overflow metabolism of sugars.

which in mammalian cells amounts to 32 ATP molecules per molecule of glucose [1].

The fermentation of sugars to ethanol by baker's yeast is a familiar example of overflow metabolism. Fermentation starts with the partial catabolism of sugars to pyruvate via glycolysis. However, instead of being further catabolised by oxidative phosphorylation, pyruvate is converted into an acid, gas or alcohol that is released from cells (overflow). The ATP yield of fermentation is that of glycolysis, 2 ATP molecules per molecule of glucose, 1/16 smaller than the ATP yield of oxidative phosphorylation.

1.2 PASTEUR, BROWN, WARBURG AND CRABTREE

In 1861, Louis Pasteur found that aeration of yeast cultures caused yeast cell growth to increase and fermentation to decrease [2]. From those observations, Pasteur postulated that fermentation takes place in anaerobic conditions – 'la vie sans gaz oxygene libre' – and that oxygen has an inhibitory effect on fermentation [3] (Pasteur effect). The Pasteur effect was later recapitulated with muscle cells [4]. When resting muscle cells were maintained in a hydrogen atmosphere (and thus anaerobic) they produced lactate at high rates. In contrast, lactate production was almost entirely abrogated in an oxygen atmosphere.

The pioneering work of Pasteur pointed to the lack of oxygen as the cause of overflow metabolism in anaerobic fermentation. Some scientist contemporaneous with Pasteur questioned his theory, that fermentation was happening at the highest rate in the absence of oxygen. Adrian J. Brown, the half-brother of the more famous Horace T. Brown, conducted experiments to measure the rate of fermentation in the absence and presence of air [5]. In those experiments, yeast was added to two flasks, one with air flow blocked and another one exposed to air flow. When he analysed the amount of alcohol on each flask 3 h pass the start of fermentation, he observed the flask exposed to air contained more alcohol. Brown's data indicated that fermentation is not restricted to anaerobic conditions. It can take place in the presence of oxygen at equal or higher rates. Brown was also the first to measure the rate of fermentation *per cell* in aerobic conditions [5]. Unfortunately, his work did not transcend his time, perhaps because of the wider acceptance of Pasteur hypothesis. I fortunately rediscovered his manuscripts when performing a Google search on the topic of aerobic fermentation. See Refs. [5,6] and a discussion of his work in Ref. [7].

It took 30 years for other biochemists to challenge the hypothesis that fermentation occurs at the highest rate in the absence of oxygen. In 1923, Otto Warburg and Seigo Minami reported that tumour tissues cultured in aerobic conditions manifested high rates of glucose fermentation to lactate, which were low in normal tissues [8]. A few years later, Crabtree reported the manifestation of aerobic fermentation by normal tissues infected with viruses [9].

The use of different experimental systems had a profound influence on the hypotheses Warburg and Crabtree pushed forward to explain aerobic fermentation. The observation of high rates of aerobic fermentation in cancer cells prompted Warburg to hypothesize that cancer cells have defective mitochondria [10,11]. That is a plausible extrapolation of the lack of oxygen as the cause of anaerobic fermentation. If fermentation occurs in the presence of oxygen, then there should be a defect in the cell machinery responsible for oxidative phosphorylation, the mitochondria in mammalian cells. In doing so, Warburg did not take into account the report by Crabtree of aerobic fermentation by normal tissues infected with viruses, which have competent mitochondria.

The recapitulation of aerobic fermentation in normal tissues subject to viral infection lead Crabtree to conclude that aerobic fermentation is not unique to cancer tissues [9]. Instead, and probably inspired by the formulation of the Pasteur effect as an inhibitory effect of oxygen on fermentation, Crabtree postulated that glycolysis has an inhibitory effect on oxidative phosphorylation: 'Evidence is brought forward which suggests that the glycolytic activity of tumours exerts a checking effect on their respiration' [12]. In doing so, Crabtree did not take into account the possibility that viruses could inhibit the mitochondrial activity of infected cells [13]. And even in the event that regulatory mechanisms mediate the inhibition of oxidative phosphorylation by glycolysis, the Crabtree hypothesis does not explain why those regulatory mechanisms were selected during evolution. Finally, both Warburg and Crabtree were unaware of the work of AJ Brown on aerobic fermentation by yeast. In honour of Brown, Warburg and Crabtree, I suggest to call aerobic fermentation the *BWC effect*.

1.3 FROM YEAST TO MARATHON RUNNERS

For historical reasons, aerobic fermentation goes under the name of 'The Warburg effect' in the context of cancer cell metabolism. The same phenomenon has gone under different names in other fields. In 1948, 20 years after the original observations by Warburg and Crabtree, Swanson published the second wave of data on aerobic fermentation by yeast cell cultures [14], presumably without knowledge of the earlier report by AJ Brown [5]. Twenty more years later, De Deken coined the term 'The Crabtree effect' as a synonym of aerobic fermentation in yeast [15]. De Deken, thinking on the lines of Crabtree, assumed that aerobic fermentation follows from an inhibition of respiration by glycolysis. From then on aerobic fermentation has gone under the name of the Crabtree effect in the context of yeast metabolism, although the original observations by Crabtree were made using mammalian cells and no definitive evidence was provided of actual inhibition of respiration by glycolysis.

By the 1930s, it was evident that lactate blood levels increase dramatically after intense physical activity relative to the lactate blood levels at rest. This observation prompted Owles W. Harlind to perform a set of controlled experiments where one subject walked for about 30 min at a predefined speed [16]. He reported: 'A critical metabolic level was found below which there was no increase in blood lactate as a result of the exercise, although above this level such an increase did occur'. By the 1930s it was also well established that lactate was the product of fermentation by muscle cells [17]. The association between fermentation and anaerobic metabolism lead to the later use of *anaerobic threshold* to denominate the threshold exercise intensity leading to blood lactate accumulation [18].

Aerobic fermentation is ubiquitous among bacteria, where it goes under the more common name of overflow metabolism. A review by JW Foster in 1947 highlights early literature with reports of overflow metabolism in bacteria [19]. Overflow metabolism in bacteria was later attributed to a limitation in the oxidative phosphorylation capacity [20–22]. Although this was the same hypothesis introduced earlier by Warburg in the context of cancer metabolism, there was no attempt to establish a connection between the observations in bacteria and mammalian cells.

The lack of cross-reference between different areas is best exemplified by the observation of aerobic fermentation in plants. After a decade of research on aerobic fermentation in bacteria, yeast and cancer, I never came across a reference to this phenomenon in plants. It was only during the preparation of this book that I became curious about the widespread occurrence of aerobic fermentation or overflow metabolism. I found two publications from Kuhlemeier lab reporting the observation of aerobic fermentation by tobacco pollen cells during development [23,24].

Table 1.1 Reports of Aerobic Fermentation Across the Life Kingdoms

	Kingdom	Organism	Major Product	Ref.
Prokaryotes	Archaea	?	?	?
	Bacteria	*Escherichia coli*	Acetate	[25]
		Aerobacter aerogenes	Acetate	[26]
		Klebsiella aerogenes	Acetate	[27]
		Selenomonas ruminantium	Lactate	[28]
		Streptococcus bovis	Lactate	[28]
Eukaryotes	Protista	Trypanosomatids	Acetate/Ethanol	[29]
	Fungi	Yeasts	Ethanol	[5]
	Plantae	Tobacco pollen	Ethanol	[23,24]
	Animalia	Cancer cells	Lactate	[8]
		Viral infections	Lactate	[9]
		Muscle cells	Lactate	[16]

(?) The search for examples in Archaea is still open.

Overflow metabolism of sugars in aerobic conditions (aerobic fermentation) has been observed across all arms of the tree of life (Table 1.1). The by-product of fermentation may differ from organism to organism, but the phenomenon is in essence the same. When sugars are provided in excess, when metabolism is pushed to its limits, cells exhibit fermentation even in aerobic conditions. The observation of aerobic fermentation came as a big surprise. It was and it is considered an inefficient form of metabolism because fermentation has a lower ATP yield than oxidative phosphorylation. It was and it is considered a wasteful phenomenon because fermentation results in overflow metabolism. It has received a plethora of names, including overflow metabolism, aerobic fermentation, aerobic glycolysis, Warburg effect, Crabtree effect and the lactate switch. There is no consensus among the scientific community regarding the cause of overflow metabolism in aerobic fermentation. Many after Brown, Warburg and Crabtree worked on the topic, but their explanations are in essence reformulations of the Warburg hypothesis: overflow metabolism is attributed to limited oxidative phosphorylation capacity. That lead us to next obvious question: What limits the oxidative phosphorylation capacity?

CHAPTER 2

Biochemical Horsepower

Everything must be made as simple as possible. But not simpler

Albert Einstein

2.1 MAXIMUM RATE OF A BIOCHEMICAL REACTION

Efficiency is a matter of context. What represents a good measure of efficiency will therefore depend on that context. We often assume *yield*, the metabolic output per unit of nutrient consumed, is the right measure of efficiency. When we said oxidative phosphorylation is more efficient than fermentation, we meant oxidative phosphorylation has a higher yield of ATP per molecule of glucose than fermentation. But yield, as a measure of efficiency, is only relevant in a context where glucose supply is limited. In an environment plenty of glucose oxidative phosphorylation does not give any advantage over fermentation. In the absence of any further evidence, we would conclude that, in an environment with plenty of glucose, oxidative phosphorylation and fermentation should be utilized by cells indistinctly. What else besides nutrients could limit metabolism?

A car analogy is a good start for inspiration. The gas pedal plays the role of increased sugar availability. How fast can a car go? At first, the car speed will depend on how far down you push the gas pedal. But, when the gas pedal is at the bottom, the car speed is determined by its engine horsepower. What is the equivalent of 'horsepower' in biochemistry?

To answer this question, let us focus on a one reaction model of a cell. A substrate or nutrient S is converted to a product P by a single biochemical reaction catalysed by enzyme E

$$S \xrightarrow{\quad E \quad} P \tag{2.1}$$

In this simple model, the metabolic rate is quantified by the Michaelis−Menten relation between the speed of a biochemical reaction

Overflow Metabolism. DOI: https://doi.org/10.1016/B978-0-12-812208-2.00002-0

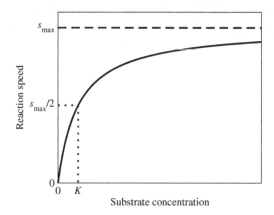

Figure 2.1 Speed of a biochemical reaction as a function of the substrate concentration. K is the half saturation constant and s_{max} is the maximum asymptotic speed when the substrate concentration is much larger than K.

and the substrate concentration (Fig. 2.1). We are interested in the maximum reaction speed when the substrate is in excess (gas pedal at the bottom). The maximum reaction speed, the plateau in Fig. 2.1, is given by enzyme turnover number times the enzyme concentration [1]

$$s_{max} = k[E] \tag{2.2}$$

The turnover number k quantifies the maximum output of one enzyme molecule: how many molecules of substrate are transformed into product by a single enzyme molecule per unit of time. The other factor is the enzyme concentration $[E]$.

In biochemistry textbooks and the metabolic modelling literature, the enzyme concentration is assumed to be a parameter set by the cell type and the cell environment. We shall go further and ask the next obvious question: what is the reaction maximum velocity when the enzyme concentration is at its maximum? To deal with this question, in line with the car analogy, I will introduce the concept of biochemical horsepower:

The *Biochemical horsepower* (*H*) is the maximum metabolic rate when the enzyme concentrations are at their maximum values.

For a single biochemical reaction the horsepower is simply the reaction maximum speed

$$H = s_{max} = k[E]_{max} \tag{2.3}$$

which is basically proportional to the maximum enzyme concentration. There are two factors determining the maximum concentration of an enzyme: size and maximum packing density. Size is quite familiar to us. We can fit more lemons than melons into a box. Both have a spherical shape, but lemons are smaller. While enzymes may look invisible from our macroscopic world, they are huge macromolecules in the nanoscopic world. Just to put things into perspective, the ribosome (20 nm) looks the size of the earth to the hydrogen atom (0.05 nm). How many enzymes can be pack in a given volume is inversely proportional to the enzyme volume (v).

The other factor, maximum packing density, depends on the object shape and the packing procedure (discussed in the next section). Ellipsoids can be packed to a higher density than spheres. Maximum theoretical packing of oranges may be common display at the grocery store, but random packing with lower density is a better model of macromolecular packing in cells. How many enzymes can be packed in a given volume is proportional to the enzyme maximum packing density for the predefined packing procedure (ϕ_{max}).

Putting these two factors together, we obtain the maximum enzyme concentration

$$[E]_{max} = \frac{\phi_{max}}{v} \tag{2.4}$$

and the horsepower for a cell model with a single biochemical reaction

$$H = \frac{k}{v}\phi_{max} \tag{2.5}$$

The horsepower of a biochemical reaction is therefore given by the enzyme turnover number per unit of enzyme volume times the maximum packing density. While the maximum packing density depends on enzyme intrinsic properties (shape, compressibility, etc.), it also depends on extrinsic factors (solvent, packing procedure, presence of other macromolecules). To exclude the contribution of extrinsic factors, I will introduce the specific horsepower:

The *Biochemical specific horsepower* (h) is the biochemical horsepower divided by the maximum packing density.

For a single biochemical reaction, the specific horsepower is

$$h = \frac{H}{\phi_{max}} = \frac{k}{v} \qquad (2.6)$$

The specific horsepower of a biochemical reaction is therefore given by the enzyme turnover number per unit of enzyme volume.

The concept of specific horsepower can be extrapolated to pathways, organelles, cells, tissues and organisms. For example, in eukaryotes, the oxidative phosphorylation machinery is located to the mitochondria. This fact becomes handy to estimate the oxidative phosphorylation specific horsepower, basically the mitochondria ATP production per unit of mitochondria volume. The specific horsepower of fermentation can be estimated from experiments reporting the rate of lactate release and the concentration of glycolysis enzymes.

Table 2.1 reports specific horsepower values for oxidative phosphorylation and fermentation [38]. Fermentation exhibits the highest value − 91 mol ATP/L/h − while the highest value for oxidative phosphorylation is about 21 mol ATP/L/h. It is the time to make a pause and reflect. We have just found an explanation for the use of fermentation at high metabolic rates: fermentation has higher horsepower than oxidative phosphorylation. A Ferrari will be a good choice for speeding in the German autobahn. It has a higher horsepower (660 vs 200 horsepower), even though it is less efficient when it comes to fuel consumption than the many Volkswagen Golfs left behind (17 vs 60 miles

Table 2.1 Specific Horsepower of Oxidative Phosphorylation and Glycolysis				
	Cell Type	Cell/Tissue	Specific Horsepower (mol ATP/L/h)	Ref.
Oxidative phosphorylation	Cancer	PC-3	2.6	[30]
		HeLa	3.0	[31]
	Liver	Liver	4.2−5.8	[32]
	Muscle	Lateralis	8	[33]
		Heart	9−12	[34]
		Gastrocnemius	12	[35]
		Soleus	12	[32]
		Plantaris	21	[32]
	Fungi	Yeast	17	[36]
Fermentation	Muscle	Glycolysis	91	[37]

per gallon). Fermentation (glycolysis) is the Ferrari of energy metabolism.

The values for oxidative phosphorylation spread over a broad range, from as low as 3 mol ATP/L/h in cancer cells to as high as 21 mol/ATP/L/h in muscle cells. The dispersion in these values is most likely a consequence of the mitochondria status in the different cell types. It is quite remarkable that the lowest values are observed for cancer cells. So Warburg may have been right after all. The mitochondria of cancer cells have lower horsepower than the mitochondria of cells in normal tissues. Defective mitochondria, as proposed by Warburg, can be reinterpreted as mitochondria with reduced horsepower.

2.2 OPTIMAL PACKING AND MAXIMUM HORSEPOWER

The specific horsepower allow us to compare different reactions and pathways based on their maximal rates per unit of enzyme, pathway enzymes or mitochondrial volume. Yet, when it comes to estimate the horsepower of a cell or tissue we need to take into consideration what fraction of the cell or tissue volume can be actually occupied by enzymes and mitochondria. For example, to calculate the eukaryote cell horsepower for energy generation via oxidative phosphorylation, we need an estimate of the maximum volume fraction ϕ_{max} that can be occupied by mitochondria. This observation leads us to problems of optimal packing [39].

Problems of optimal packing have fascinated scientists since antiquity. The most celebrated moment was the conjecture by Johannes Kepler in 1611. Kepler proposed that the densest possible packing of spheres is a face-centred cubic packing, occupying a volume fraction of $\pi/\sqrt{18} \approx 0.74$ [39]. It took centuries to obtain a mathematical proof of this conjecture. Between the 1990s and 2000s the Mathematician Thomas C. Hales reported his mathematical proof, which 'can be read' in a 120-page manuscript published by Annals of Mathematics [40]. Hales mathematical proof is so intricate that it is still under scrutiny.

The face-centred cubic arrangement is the densest packing, but we do not expect such arrangement to be a good model of proteins and organelles within cells. Perhaps a better representation is random packing. When it comes to random packing we need to specify the mechanism of how randomness is generated. If packing is enhanced through vibrations then we achieve what is known as the *closest random packing* of spheres, occupying a volume fraction of about 0.64 [41,42].

The requirement of vibrations to attain the closest random packing makes it unsuitable to model mitochondria packing. Mitochondria are large organelles with reduced motility within cells. And mitochondria do not possess a spherical shape in general. Based on a review of the literature, I find the packing by random sequential addition of spheroids the closest model of mitochondria packing within a cell. In the *random sequential addition* procedure we add objects into a given volume. Whenever the object does not overlap with previously allocated objects the new object is accepted, otherwise the placement is rejected and we try a new allocation. Fig. 2.2 shows the volume fractions obtained by random sequential addition of spheroids as a function of the spheroid aspect ratio, based on computer simulations [43]. The best packing is obtained for a slightly spheroidal shape, about two times longer than wider, with a volume fraction just above 0.4.

The other factor we need to take into consideration is the formation of mitochondria aggregates, changing the nature of the problem to packing of aggregates. The packing of rigid aggregates subject to weak compaction forces reaches a volume fraction of 0.36 independently of

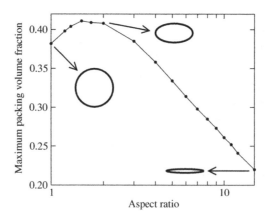

Figure 2.2 Maximum volume fraction occupied by spheroids packed by random sequential addition. Based on computer simulations data (Sherwood JD. Packing of spheroids in three-dimensional space by random sequential addition. J Phys a-Math Gen 1997;30(24):L839–43).

the spatial dimension of the aggregates [44]. Therefore, if we take random sequential addition or packing of rigid aggregates as potential representations for the optimal packing of mitochondria in a cell, then we conclude that mitochondria can occupy a maximum volume fraction in the range 0.36−0.41.

Table 2.2 reports measurements of the mitochondrial volume fraction in cells of different tissues. The maximum values are reported for cardiac muscle and the flight muscle of Hummingbirds, in the range 0.35−0.38, within the theoretical estimate based on geometrical considerations. In other words, the mitochondrial content of cardiac muscle and Hummingbirds flight muscle are at their maximum theoretical values based on packing constraints.

Through this book, we will assume a maximum packing density of 0.4 for the cell volume, which is about what predicted by random sequential addition. When it comes to estimate content per unit of tissue volume we should also take into account that cells do not fill all the space in a tissue. We can estimate the maximum tissue volume fraction occupied by mitochondria from the product of the tissue volume fraction occupied by cells times the maximum cell volume fraction occupied by mitochondria (about 0.4). This procedure is applied in Table 2.3 using as input reports of cell density within a tumour and skeletal muscle, resulting in values around 30%. These estimates are

Table 2.2 Mitochondrial Volume Fraction Across Tissues			
Tissue	Name/Description	ϕ_M	Ref
Cancer	HeLa cells	0.07	[45]
	Osteosarcoma cells	0.077	[46]
	Rat hepatoma cell line	0.06−0.10	[47]
	PC-12 cells	0.1	[48]
Brain	Monkey brain cells	0.07	[49]
	Cortical neurons	0.1	[50]
Kidney	Rat renal tubular epithelium	0.14	[51]
Muscle	Hamster muscle, glycolytic	0.1	[52]
	Hamster muscle, oxidative	0.2	[52]
	Cardiac muscle myofilaments, human	0.25	[53]
	Cardiac muscle myofilaments, mouse	0.38	[53]
	Fish muscles	0.16−0.37	[54,55]
	Hummingbird flight muscles	0.35	[56]

Table 2.3 Mitochondria Volume Fraction Within Tissues

Tissue	Name/Description	$\phi_{cells,\ tissue}$	$\phi_{M,\ tissue}$	Ref.
Tumour	MDA PCa 2b subcutaneous tumour	0.72	0.29	[59]
Muscle	Skeletal muscle	0.8–0.85	0.32–0.34	[59]

consistent with the fact that water represents 75% of tissue weight [57,58]. Taking the water density of 1 kg/L that will correspond to 75% of tissue volume. The remaining 25% is attributed to insoluble material including between mitochondria and other macromolecules. Our estimates based on cell density and maximum intracellular mitochondrial density are closer to 30%, just slightly above the expected 25% based on muscle water content. Through this book, we will assume a maximum packing density of 0.25 for the tissue volume.

CHAPTER 3

How Fast Can We Run?

When you have eliminated the impossible, whatever remains, however improbable, must be the truth

Shelock Holmes

3.1 THE LACTATE THRESHOLD

Breaking the Marathon world record is all about pushing your running pace to its limit. Yet, we cannot run a Marathon race at a 100-m race pace. To find out your long distance running pace you can try the following experiment. Jump on a treadmill and set it to a slow walking pace. After 5 min, increase the speed to a slow jogging pace. Your breathing rate will increase but you will be able to keep that pace for some time. Keep doing that, increase the running speed by a small amount and keep that speed for 5 min. No matter whether you have been running regularly or not at all, you will find a speed that you cannot sustain for the next 5 min. That speed is your long distance running pace, which is known in sport medicine as the anaerobic threshold or lactate threshold speed.

The *lactate threshold* owes its name to the physiological changes occurring when we exceed the threshold exercise intensity. By the 1930s, it was evident that lactate blood levels increase dramatically after intense physical activity, relative to the lactate blood levels at rest. This observation prompted Owles W. Hardind to perform a set of controlled experiments where one subject walked for about 30 min at a predefined speed [16]. He reported: 'A critical metabolic level was found below which there was no increase in blood lactate as a result of the exercise, although above this level such an increase did occur'. By the 1930s, it was also well established that lactate was the product of fermentation by muscle cells [17]. The association between fermentation and anaerobic metabolism lead to the later use of *anaerobic threshold* to denominate the threshold exercise intensity leading to blood lactate accumulation [18].

Overflow Metabolism. DOI: https://doi.org/10.1016/B978-0-12-812208-2.00003-2

The lactate threshold demarks a transition from oxidative phosphorylation to mixed oxidative phosphorylation and fermentation with lactate overflow. One way to interpret this transition is to say that oxidative phosphorylation has a limited capacity for energy generation. Once the exercise intensity requires an energy demand exceeding the oxidative phosphorylation capacity, fermentation satisfies the excess energy demand. But what determines the oxidative phosphorylation capacity of muscle tissue?

Right at the onset of the lactate threshold we experience an increase in the breathing rate. It feels as if we are running out of oxygen. This evidence suggests a link between the maximum metabolic capacity of our muscles and our aerobic capacity. The *aerobic capacity*, or $\dot{V}_{O_2,max}$, is defined as the maximum amount of oxygen that a subject can use per unit of time and body weight [60]. The aerobic capacity is typically reported in units of mL O_2/min/kg of body weight. Our experience running out of oxygen after intense physical activity would suggest that our aerobic capacity is determined by factors affecting the oxygen supply to the muscle cells. That includes the lungs capacity, heart strength, concentration of haemoglobin in blood, and the capillary density of muscle tissue. Such a common assumption blinds us from considering the possibility that our aerobic capacity matches the intrinsic body capacity for aerobic metabolism. By that I mean that our body has experienced a certain maximum demand of aerobic metabolism and our lungs and other factors determining our aerobic capacity have been adjusted to satisfy the associated oxygen demand. I will hypothesize that we can determine the limits of muscle metabolism under the assumption that the oxygen supply matches the demand of aerobic metabolism.

3.2 THE ENERGY COST OF RUNNING

The energy demand associated with a physical activity can be estimated from the increase in oxygen consumption above rest $\Delta \dot{V}_{O_2}$, generally reported in units of litre/hour/kg of body weight [61,62]. Assuming that the increase in oxygen consumption matches the increase in oxidative phosphorylation we obtain an estimate of the energy demand associated with the physical activity (mol ATP/h)

$$J_E = 2P_O \frac{\Delta \dot{V}_{O_2}}{V_{STP}} \tag{3.1}$$

where P_O is the P/O ratio and V_{STP} is the gas molar volume at standard temperature and pressure and. $2P_O$ is introduced to convert from moles of O_2 consumed to moles of ATP produced by oxidative phosphorylation. V_{STP} is introduced to convert the oxygen consumption rate from the typical units of litre/hour to mol/hour. In the context of running, the oxygen consumption rate above rest is a function of the running speed (s) and body weight [61,62]

$$\Delta \dot{V}_{O_2} = as \left(\frac{W}{W_0} \right)^{-0.4} \tag{3.2}$$

where $a = 8.5$ L/km/kg of body weight and $W_0 = 0.001$ kg of body weight. Combining Eqs. (3.1) and (3.2), we obtain a working relationship between the energy demand and running speed

$$J_E = \varepsilon s \tag{3.3}$$

where

$$\varepsilon = \frac{2P_O \Delta \dot{V}_{O_2}}{V_{STP} s} = a \left(\frac{W}{W_0} \right)^{-0.4} \frac{2P_O}{V_{STP}} \tag{3.4}$$

is the energy demand of running per unit of speed and body weight. For typical body weights between 50 and 100 kg, the energy demand per unit of speed and body weight (assuming $V_{STP} = 22.4$ L/mol and $P_O = 2.5$) varies between 25 and 19 mmol ATP/km/kg of body weight, respectively. In the following, we will assume the typical value

$$\varepsilon \sim 22 \text{ mmol ATP/km/kg of body weight} \tag{3.5}$$

corresponding to a body weight of 70 kg.

J_E gives us an estimate of the energy demand as deduced from the oxygen consumption rate above rest. Yet, some distinction should be made regarding the specific tissue that increases its rate of oxidative phosphorylation. And that will depend on whether the energy demand at the exercising muscle is met by aerobic or fermentation metabolism.

3.3 RUNNING WITH AEROBIC METABOLISM

We will start with the hypothetical case where all the energy required for running is generated by oxidative phosphorylation at the exercising muscle. I will define aerobic metabolism as the physiological context where the energy generation and the oxygen consumption rate above resting are both determined by mitochondrial oxidative phosphorylation at the exercising muscle.

In aerobic metabolism, the tissue responsible for the increase in oxygen consumption is the exercising muscle itself. The running energy demand deduced from the oxygen consumption above rest (J_E) matches the energy generation by aerobic metabolism (r_{AM}) at the exercising muscle

$$J_E = r_{AM} = v_E \phi_{EM} h_M \tag{3.6}$$

where v_E is the volume of exercising muscle per unit of body weight, ϕ_{EM} is the volume fraction of exercising muscle occupied by mitochondria and h_M is the mitochondria specific horsepower for energy generation. From this equation, we can estimate the mitochondria content necessary to maintain a speed s using AM.

$$\phi_{EM} = \frac{J_E}{v_E h_M} = \frac{\varepsilon}{v_E h_M} s \tag{3.7}$$

Table 3.1 reports the estimated mitochondria volume fraction in lower body skeletal muscle (LBSM) that would be required to make the world records for different running distances. For long distance races above 10 km mitochondria are estimated to occupy about 20% of the LBSM volume. To put this number into perspective let's take a look at the typical muscle composition. Water represents 75% of the muscle weight [57,58]. Assuming a density of 1 kg/L that will be 75% of the muscle volume. The remaining 25% is distributed between mitochondria, myofibrils and other components. The myofibrils transduce the energy produced by mitochondria into mechanical work and they are essential for muscle function. So a mitochondrial content of 20% of LBSM volume is quite high, just 5% short the 25% cap.

For short distance races below 1500 m, mitochondria are estimated to exceed 25% of the LBSM volume. Of course, this is just a though experiment, a *Gedanken experiment* as Einstein would say. Based on the reported muscle composition, the mitochondria content does not exceed 25% of the LBSM volume. Therefore, we are force to conclude that mitochondrial oxidative phosphorylation cannot sustain the intense muscle activity to break the world record of races below 1500 m.

At this point, let's summarize what we have achieved. Even under the assumption that oxygen supply is in excess, mitochondria cannot

Table 3.1 Exercising Muscle Mitochondrial Volume Fraction (ϕ_{EM}) Needed to Match the World Record Time for Men Outdoor Races using Aerobic Metabolism at the Exercising Muscle

Discipline	Time	Competitor	s (km/h)	ϕ_{EM} (%)
100 m	9.58 s	Usain Bolt	37.6	34[a]
200 m	19.19 s	Usain Bolt	37.5	34[a]
400 m	43.03 s	Wayde van Niekerk	33.5	30[a]
800 m	00:01:41	David Lekuta Rudisha	28.5	26[a]
1000 m	00:02:12	Noah Ngeny	27.3	24
1500 m	00:03:26	Hicham El Guerrouj	24.2	22
2000 m	00:04:45	Hicham El Guerrouj	25.3	23
3000 m	00:07:21	Daniel Komen	24.5	22
5000 m	00:12:37	Kenenisa Bekele	23.8	21
10,000 m	00:26:18	Kenenisa Bekele	22.8	20
10 km	02:44:00	Leonard Patrick Komon	22.4	20
15 km	17:13:00	Leonard Patrick Komon	21.8	20
20,000 m	00:56:26	Haile Gebrselassie	21.3	19
20 km	07:21:00	Zersenay Tadese	21.7	19
25,000 m	01:12:25	Moses Cheruiyot Mosop	20.7	19
25 km	01:11:18	Dennis Kipruto Kimetto	21.0	19
30,000 m	01:26:47	Moses Cheruiyot Mosop	20.7	19
30 km	01:27:13	Eliud Kipchoge	20.6	18
30 km	01:27:13	Stanley Kipleting Biwott	20.6	18
Marathon	02:02:57	Dennis Kipruto Kimetto	20.5	18
100 km	06:13:33	Takahiro Sunada	16.1	14

Assuming a lower body skeletal muscle specific volume of 0.21 L/kg [63] and the median mitochondrial specific horsepower of muscle cells (Table 2.1). The record times and athletes were obtained from the IAAF website on January 2017.
[a]*The required mitochondrial volume fraction is not feasible.*

satisfy the energy requirements of intense physical activity. This is due to the existence of a volume fraction cap and the limited specific horsepower of mitochondria. What about fermentation?

3.4 RUNNING WITH FERMENTATION METABOLISM

We will now consider the hypothetical case where all the energy required for running is generated by fermentation. When it comes to fermentation there are two issues we need to deal with. One is, as for AM, the energy generation at the exercising muscle. Second, since fermentation release lactate, some other tissue needs to take care of turning over lactate. One way to do so is through the Cori cycle [1]

$$\text{glucose} + 2\,\text{ADP} \quad \rightarrow \quad \text{lactate} + 2\,\text{ATP} \quad \text{Exercising muscle}$$
$$\text{lactate} + 3\,\text{ATP} \quad \rightarrow \quad \text{glucose} + 3\,\text{ADP} \quad \text{Peripheral tissue}$$

where lactate is turnover into glucose by gluconeogenesis in peripheral tissue. On the original work of Cori the liver was proposed as the peripheral tissue. Nowadays it is recognized that other tissues can turn-over lactate as well. The Cori cycle operates at expenses of an energy burden in the peripheral tissue. The energy production rate at the peripheral tissue is 3/2 times the rate of fermentation at the exercising muscle. This energy demand is satisfied by oxidative phosphorylation in the peripheral tissue. Since glucose is being regenerated to sustain fermentation at exercising muscles, oxidative phosphorylation is fuelled by other substrates, generally fatty acids.

Taking these facts into consideration I will define *fermentation metabolism* as the steady state physiological context where (1) fermentation satisfy the energy demand of physical activity and (2) oxidative phosphorylation satisfy the energy demand of turning over lactate.

In fermentation metabolism the tissue responsible for the increase in oxygen consumption is the peripheral tissue. Therefore the running energy demand deduced from the oxygen consumption above rest (J_E) is 3/2 times the rate of fermentation and it matches the energy generation by mitochondria at the peripheral tissue

$$J_E = \frac{3}{2} r_F = v_P \phi_{\text{PM}} h_M \tag{3.8}$$

where v_P is the specific volume of peripheral tissue per unit of body weight and ϕ_{PM} is the peripheral tissue volume fraction occupied by mitochondria. The energy generation by fermentation at the exercising muscle is matched by the corresponding content of fermentation enzymes

$$r_F = v_E \phi_{\text{EF}} h_F \tag{3.9}$$

where v_E is the specific volume of exercising muscle, ϕ_{EF} is the exercising muscle volume fraction occupied by fermentation enzymes and h_F is the horsepower of fermentation for energy generation. From the equations above we can estimate both the volume fraction of fermentation enzymes at exercising muscle and the mitochondria volume fraction at peripheral tissues that are needed to sustain the fermentation mode of muscle metabolism.

$$\phi_{\text{EF}} = \frac{2}{3}\frac{\varepsilon}{v_E h_F}s \tag{3.10}$$

$$\phi_{\text{PM}} = \frac{\varepsilon}{v_P h_M}s \tag{3.11}$$

The estimated fermentation enzymes fraction in LBSM that would be required to make the world records for different distances does not exceed 1% for typical races, far below the 25% cap of biomass composition of muscle tissue. This result indicates that space availability, crowding, of fermentation enzymes at the exercising muscle is not a limitation. Fermentation has such a high horsepower that a small volume percentage is sufficient to sustain the energy demand of running. Yet, there is a limit on the peripheral tissue capacity to turnover lactate.

Table 3.2 reports the estimated mitochondria volume fraction in peripheral tissue that is required to clear the lactate generated at the exercising muscle. Since a priori it is not obvious what tissue could be carrying on that function, the table reports the values for different assumptions of peripheral tissue. If the liver represents the peripheral tissue then it will require mitochondria volume fractions above 100%, independently of the speed. Therefore, based on these calculations, the liver alone cannot sustain the Cori cycle during running at world record speed, independently of the race distance. If the upper body skeletal muscle represents the peripheral tissue then it will generally require mitochondria volume fractions above 25%. Since that exceeds the 25% cap of biomass composition of muscle tissue, the upper skeletal muscle mass is neither sufficient to sustain the Cori cycle during running at world record speed. Only when we assume that all skeletal muscle represents the peripheral tissue we obtain mitochondria volume fractions below 25%.

3.5 THE IMPACT OF NUTRIENT STORAGE

So far we have not taken into consideration the nutrient requirements of running. We can estimate the minimum nutrient reserves that are required to run a race taking into account the typical energy demand per unit of running speed (ε, mol ATP/km/kg of body weight) and the

Table 3.2 Peripheral Muscle Mitochondrial Volume Fraction (ϕ_{PM}) Needed to Match the World Record Time for Men Outdoor Races Using the Cori Cycle

Discipline	Time	Competitor	s (km/h)	ϕ_{PM} (%) Liver	UBSM	SM
100 m	9.58 s	Usain Bolt	37.6	354[a]	44[a]	19
200 m	19.19 s	Usain Bolt	37.5	353[a]	44[a]	19
400 m	43.03 s	Wayde van Niekerk	33.5	315[a]	39[a]	17
800 m	00:01:41	David Lekuta Rudisha	28.5	269[a]	34[a]	14
1000 m	00:02:12	Noah Ngeny	27.3	257[a]	32[a]	14
1500 m	00:03:26	Hicham El Guerrouj	24.2	228[a]	28[a]	12
2000 m	00:04:45	Hicham El Guerrouj	25.3	238[a]	30[a]	13
3000 m	00:07:21	Daniel Komen	24.5	231[a]	29[a]	12
5000 m	00:12:37	Kenenisa Bekele	23.8	224[a]	28[a]	12
10,000 m	00:26:18	Kenenisa Bekele	22.8	215[a]	27[a]	11
10 km	02:44:00	Leonard Patrick Komon	22.4	211[a]	26[a]	11
15 km	17:13:00	Leonard Patrick Komon	21.8	205[a]	26[a]	11
20,000 m	00:56:26	Haile Gebrselassie	21.3	200[a]	25[a]	11
20 km	07:21:00	Zersenay Tadese	21.7	204[a]	26[a]	11
25,000 m	01:12:25	Moses Cheruiyot Mosop	20.7	195[a]	24[a]	10
25 km	01:11:18	Dennis Kipruto Kimetto	21.0	198[a]	25[a]	10
30,000 m	01:26:47	Moses Cheruiyot Mosop	20.7	195[a]	24	10
30 km	01:27:13	Eliud Kipchoge	20.6	194[a]	24	10
30 km	01:27:13	Stanley Kipleting Biwott	20.6	194[a]	24	10
Marathon	02:02:57	Dennis Kipruto Kimetto	20.5	193[a]	24	10
100 km	06:13:33	Takahiro Sunada	16.1	151[a]	19	8

Assuming the median mitochondrial specific horsepower of muscle cells (Table 2.1). Three cases where the peripheral tissue was represented by the liver (specific volume 0.02 L/kg), upper body skeletal muscle (UBSM, specific volume of 0.16 L/kg [63]) and whole body skeletal muscle (SM, specific volume of 0.38 L/kg [63]) were considered. The record times and athletes were obtained from the IAAF website on January 2017.
[a]The required mitochondrial volume fraction is not feasible.

energy yield of the chosen nutrient (γ, mol ATP/mol of nutrient). The nutrient content (C_N, mol/kg of body weight) required to run a distance d is bound by the equation

$$C_N \geq \frac{\varepsilon d}{\gamma} \qquad (3.12)$$

If we further assume that the nutrient is stored in skeletal muscle, then we can estimate the skeletal muscle volume fraction that is needed to store that amount of nutrient

$$\phi_N = C_N \frac{m_N/\rho_N}{v_{SM}} \qquad (3.13)$$

where v_{SM} is the skeletal muscle specific volume (L/kg of body weight), m_N is the nutrient molecular weight (g/mol) and ρ_N is the nutrient density (kg/L). Combining Eqs. (3.12) and (3.13), we obtain

$$\phi_N \geq \frac{\varepsilon\left(m_N/\rho_N\right)}{\gamma v_{SM}} d \qquad (3.14)$$

Table 3.3 reports the percentage of skeletal muscle volume, as estimated from the equation above, that is required to allocate the minimum nutrient demand per km, and the total amount required to run a Marathon race. If the choice of nutrient storage is glycogen then the runner will require at least 2% of its skeletal muscle to be occupied by glycogen to finish the Marathon race. This number goes down to about 1/2% if palmitic acid is chosen instead. These estimates indicate that nutrient storage is not a big burden for races under 42 km. Therefore, we can conclude that nutrient storage is not a key factor to determine the limits of muscle metabolism in the context of physical activities lasting a few hours.

The opposite is true for extreme physical activities lasting several hours. For example, running a 100-km race would require our body to allocate about 5% of skeletal muscle for glycogen storage or 1.5% for palmitic acid storage. These numbers are now getting in the range where they are limiting the amount of muscle space available for the allocation of the mitochondria and myofibrils.

Table 3.3 Minimum Nutrient Storage Requirements of Running				
Energy Source	**Parameter**	**Value**	**Units**	**Reference**
Glycogen	Energy yield (G1P)	33	mol/mol	
	Molecular weight (G1P)	260	g/mol	[64]
	ϕ_N (% of skeletal muscle/km)	0.046		Eq. (3.14)
	ϕ_N (% of skeletal muscle/42 km)	1.9		Eq. (3.14)
Palmitic acid	Energy yield	106	mol/mol	[1]
	Molecular weight	256	g/mol	[64]
	ϕ_N (% of skeletal muscle/km)	0.014		Eq. (3.14)
	ϕ_N (% of skeletal muscle/42 km)	0.59		Eq. (3.14)
Assuming a nutrient density of $\rho_N = 1$ kg/L.				

3.6 THE IMPACT OF BLOOD CIRCULATION

In fermentation metabolism, the lactate generated at the exercising muscle is transported to peripheral tissues for its conversion back to glucose. This job is done by the blood circulation system. In order to remain on a balanced physiological state, the generation of lactate should be matched by the blood flow of lactate. The balance between lactate generation at the exercising muscle, lactate in blood circulation and lactate consumption at the peripheral tissue is given by the equation

$$r_F = \frac{QC_L\beta_P}{W} = \frac{2}{3}\varepsilon s \qquad (3.15)$$

where Q is the cardiac output (litre of blood/min), C_L is the concentration of lactate in blood, β_P is the fraction of blood circulation flowing through the peripheral tissue, W is the body weight and the last equality follows from Eq. (3.8). From this equation, we obtain a linear relationship between blood plasma concentration and running speed

$$C_L = \frac{2\varepsilon W}{3Q\beta_P}s \qquad (3.16)$$

with a slope determined by the apparent energy demand per unit of running speed (ε), the body weight (W), the cardiac output (Q) and the fraction of blood circulation through the peripheral tissue (β_P). When we plug in typical parameter values in this equation ($\varepsilon = 22$ mmol ATP/km/kg, $Q = 6$ L/min, $\beta_P = 15\%$, $W = 70$ kg), we obtain an astonishing value as high as 190 mM of lactate in blood when running a typical speed of 10 km/h. Just to give a reference, the typical glucose concentration in blood is around 5 mM. Relative to that number 190 mM of lactate seem completely nonphysiological. We have found the Acchiles heel of fermentation metabolism, it requires our body to tolerate massive amounts of circulating lactate. Running with fermentation metabolism is only feasible for a short period of time, before the harmful effects of high lactate levels kick in.

The lactate concentration in blood is inversely proportional to the cardiac output (Eq. (3.16)). Therefore, a mechanism to reduce lactate levels in blood is to increase the cardiac output, by increasing the cardiac rate for example. This observation provides a hypothesis for the increased cardiac rate once the lactate threshold has been crossed.

Bear in mind that the increase in cardiac output during intense physical activity is generally attributed to the oxygen demand of aerobic metabolism. Our analysis demonstrates that, even when oxygen is available in excess, a high cardiac output is needed to transport the lactate generated at the exercising muscle to the peripheral tissue for its turnover. The latter implies that high blood lactate and increased heart beat should not be taken as definitive evidence that there is a deficient oxygen supply to muscle.

3.7 MIXED AEROBIC–FERMENTATION METABOLISM

From the analysis of hypothetical scenarios running with pure aerobic- or fermentation-metabolism, we arrive to the following observations. Running with pure aerobic metabolism we cannot break the world record of races around or shorter than 1000 m. It can be done with fermentation metabolism but at expense of the accumulation of lactate in blood. To run fast for an extended period of time, we must find a compromise between aerobic and fermentation metabolism. In the following, we will assume that, for a given running speed, the 'optimal' solution is to run with the combination of aerobic and fermentation metabolism that minimizes the concentration of lactate in blood. To uncover the optimal compromise between these two pathways, we will make the following assumptions:

1. In mixed aerobic/fermentation metabolism, there is a contribution from both the exercising muscle and the peripheral tissue to the increase in oxygen consumption above rest

$$J_E = r_{AM} + \frac{3}{2} r_{FM} = \varepsilon s \qquad (3.17)$$

where r_{AM} and r_F are the rates of aerobic and fermentation metabolism at the exercising muscle per unit of body weight.

2. The rates of aerobic and fermentation metabolism are determined by the corresponding horsepower and volume fractions of mitochondria and fermentation enzymes

$$r_{AM} = v_E \phi_{EM} h_M \qquad (3.18)$$

$$r_F = v_E \phi_{EF} h_F \qquad (3.19)$$

3. As in the case of pure fermentation, the lactate production and consumption are balanced through the Cori cycle, whereby the energy demand of gluconeogenesis from lactate ($3/2 \, r_F$) is sustained by aerobic metabolism at peripheral tissues

$$\frac{3}{2} r_F = v_P \phi_{PM} h_M \tag{3.20}$$

4. The blood lactate concentration associated with the fermentation component, as obtained from Eq. (3.15), is given by

$$C_L = \frac{W}{Q\beta_P} r_F \tag{3.21}$$

5. The overall volume fraction of mitochondria and fermentation enzymes at the exercising muscle and of mitochondria at peripheral tissues cannot exceed the maximum allowed volume fraction ϕ_{max}

$$\phi_{EM} + \phi_{EF} \le \phi_{max}$$
$$\phi_{PM} \le \phi_{max} \tag{3.22}$$

where ϕ_{max} is about 25% [57,58].

For models (1)–(5), the optimal flux distribution (r_{AM}, r_F) that minimizes the blood concentration of lactate (Eq. (3.21)) is given by

$$r_{AM} = \begin{cases} \varepsilon s & 0 \le s < s_{max, A} \\ \varepsilon s_{max, A} & s_{max, A} \le s \le s_{max, F} \end{cases} \tag{3.23}$$

$$r_F = \begin{cases} 0 & 0 \le s < s_{max, A} \\ \dfrac{2}{3} \varepsilon \left(s - s_{max, A} \right) & s_{max, A} \le s \le s_{max, F} \end{cases} \tag{3.24}$$

$$C_L = \begin{cases} 0 & 0 \le s < s_{max, A} \\ \lambda_s \left(s - s_{max, A} \right) & s_{max, A} \le s \le s_{max, F} \end{cases} \tag{3.25}$$

$$s_{max, A} = \frac{v_E h_M}{\varepsilon} \phi_{max} \tag{3.26}$$

$$s_{max, F} = \frac{v_P h_M}{\varepsilon} \phi_{max} \tag{3.27}$$

$$\lambda_s = \frac{2\varepsilon W}{3Q\beta_P} \tag{3.28}$$

Table 3.4 Parameters of Muscle Activity and Lactate Overflow			
Parameter	Value	Units	Reference
h_M	12	mol ATP/L/h	Median of muscle tissue (Table 2.1)
v_E	0.21	L/kg of body weight	Lower body skeletal muscle [63]
v_P	0.38	L/kg of body weight	Skeletal muscle [63]
ϕ_{max}	25	% of tissue	[57,58]
ε	22	mmol ATP/km/kg	Eq. (3.5)
$s_{max,A}$	29	km/h	Eq. (3.26)
$s_{max,F}$	52	km/h	Eq. (3.27)
$\Delta \dot{V}_{O_2, max}$	85	mL/kg/min	Eq. (3.29)
$Q\beta_P$	0.75	L/min	At rest [60]
λ_s	23	mM/(km/h)	At rest, Eq. (3.28)
$Q\beta_P$	20	L/min	Heavy exercise [60]
λ_s	1	mM/(km/h)	Heavy exercise, Eq. (3.28)

When there is enough room for mitochondria at the exercising muscle ($\phi_{EM} < \phi_{max}$), the optimal solution is to run with aerobic metabolism. This holds up to a maximum speed $s_{max,A}$ given by Eq. (3.26). Above this speed, running with aerobic metabolism is not feasible and fermentation must take place. For speeds above $s_{max,A}$, the rate of fermentation metabolism increases linearly with the running speed (3.24). Consequently the concentration of lactate in blood also increases linearly with the running speed (3.25), with slope λ_s.

Table 3.4 reports quantitative estimate of the predicted lactate threshold velocity $s_{max,A}$ assuming a mitochondrial volume fraction of 25% of tissue. We obtain a threshold velocity of 29 km/h. Of course, we do not expect the mitochondria volume fraction to be equal for every individual. The amount of mitochondria in LBSM will depend on the level of training and type of training. And the predicted lactate threshold velocity is proportional to the mitochondrial volume fraction in LBSM (Fig. 3.1). It ranges from 0 to 30 km/h depending on the mitochondria volume fraction at exercising tissue.

The mixed aerobic–fermentation metabolism can continue until the running speed reaches the maximum speed that can be sustained by fermentation metabolism ($s_{max,F}$, Eq. (3.27)). The oxygen consumption at this speed

$$\Delta \dot{V}_{O_2,max} = \frac{V_{STP}}{2P_O} \varepsilon s_{max, F} \tag{3.29}$$

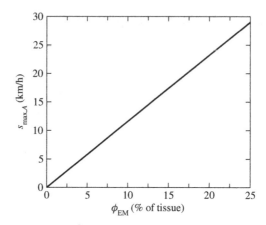

Figure 3.1 Linear dependency of the lactate threshold speed with the mitochondrial volume fraction at exercising muscle.

is basically our estimate for the maximum aerobic capacity. Table 3.4 reports the aerobic capacity deduced from this equation after plugging in parameter estimates from the literature, assuming that muscle are fully packed with mitochondria. We obtain a maximum aerobic capacity of 85 mL O_2/kg of body weight/min. This estimate is in the range of the aerobic capacity of elite male long-distance runners and crosscountry skiers (80−84 mL O_2/kg of body weight/min [60], page 128). As for the lactate threshold speed, the aerobic capacity will depend on what is that actual volume fraction of peripheral tissue that is occupied by mitochondria (Fig. 3.2). It ranges from 0 to 85 mL O_2/kg of body weight/min depending on the mitochondria volume fraction at peripheral tissue.

We can now put all together and test how the theoretical prediction given by Eqs. (3.23)−(3.28) compares with actual measurements. Fig. 3.3 shows blood lactate levels as a function of running speed. This data was obtained from experiments where subjects ran on a treadmill at a controlled speed [65]. Below the threshold velocity of about 16 km/h, the blood lactate levels are low and around the basal lactate levels at rest. When the threshold velocity of 16 km/h is exceeded, blood lactate levels increase approximately linear with the running speed. On top of the experimental points, we have overlaid the theoretical prediction assuming a mitochondrial volume fraction at the LBSM of $\phi_{EM} = 13\%$ and a muscle blood flow corresponding to the value observed during intense physical activity $Q\beta_P = 20$ L/min [60]. This is

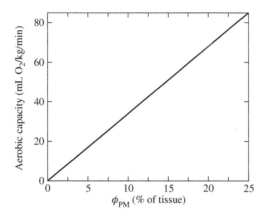

Figure 3.2 Linear dependency of the aerobic capacity with the mitochondrial volume fraction at peripheral tissue.

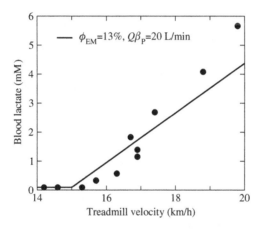

Figure 3.3 Lactate blood concentration at different running speeds. The symbols are measurements [65], and the line is the theoretical expectation for the inset parameters.

quite an impressive fit of the theoretical law to the data. It is a demonstration that the limits of physical activity can be deduced, as a first approximation, from the mitochondria horsepower and mitochondria packing density.

CHAPTER 4

How Fast Can We Grow?

Anyone who believes exponential growth can go on forever in a finite world is either a madman or an economist

Kenneth Boulding

4.1 THE RIBOSOME HORSEPOWER

In order to proliferate, cells duplicate their biomass. Once the cell doubles its initial mass, it divides producing two approximate copies of itself. Cell proliferation is therefore the sum of cell growth and subsequent cell division. In the absence of cell death, the rate of cell proliferation matches the rate of cell growth. Cell growth in turn requires both catabolic and anabolic pathways. Catabolic pathways provide energy and precursor molecules. Anabolic pathways use energy and precursor metabolites to synthesize more complex molecules that make a cell: proteins, lipids and nucleotides. Since cell growth is a combination of catabolism and anabolism, the maximum rate of cell growth will be achieved when the rates of anabolism and catabolism are optimized concomitantly. We can anticipate that aspects of both anabolism and catabolism will be relevant when determining which metabolic pathways are more efficient to sustain or maximize growth under specified conditions.

The major component of the cell biomass is protein. Proteins account for 55% of *Escherichia coli* dry weight [66], 45% of yeast dry weight [67], and 60% of the dry weight of mammalian cells [68]. Based on this observation, we will attempt to estimate the maximum rate of cell growth assuming that the cell is only composed of proteins. If amino acids and ATP are provided in excess, the cell growth rate will be limited by the speed at which a ribosome can duplicate the protein content of the cell, i.e., the ribosome horsepower for protein synthesis.

As an illustration let's estimate the maximum growth rate of *E. coli* cells. The *E. coli* ribosomes have a turnover number of $k_R = 22$ amino acids/s [69] and a volume of $v_R = 1.6$ L/mmol of ribosomes [70],

Overflow Metabolism. DOI: https://doi.org/10.1016/B978-0-12-812208-2.00004-4

resulting in a specific horsepower of $h_R = k_R/v_R = 46$ mol of amino acid/hour/litre of ribosome. Approximating the ribosomes maximum packing density by the macromolecular density of cells, $\phi_{max} \sim 0.4$ [71], we obtain a ribosome horsepower of $H_R = \phi_{max}h_R = 18$ mol of amino acid/hour/litre of cell. The ribosome horsepower represents the maximum protein synthesis rate the cell can achieve under the assumption that all the cell biomass is made of ribosomes.

The maximum growth rate can be estimated by taking into account that at maximum growth rate the protein synthesis rate is equal to $\mu_{max}P$, where P is the cell protein concentration, and it is matched by the cell horsepower for protein synthesis

$$H_R = \mu_{max}P \tag{4.1}$$

The protein concentration in a typical E. coli cell can be estimated by dividing the protein density of E. coli ($\rho_P = 0.25$ kg protein/L of cell [72]) by the average molecular weight of an amino acid ($m_a = 109$ g/mol), obtaining $P = 2.3$ mol of amino acid/L of cell. Then from Eq. (4.1), we can estimate the maximum growth that can be sustained by the ribosomes horsepower $\mu_{max} = H_R/P = 7.8$ doublings/h. The maximum proliferation rate of E. coli cells in a rich medium is about 2 doublings/h [73], about one-fourth of the maximum growth rate estimated from the ribosome horsepower. Not bad for a start.

4.2 MINIMAL MODEL OF CELL GROWTH

To obtain a better estimate of the cell maximum growth rate, we bring in other important factors, such as the energy producing machinery and structural proteins present in the cell for functions not directly linked to metabolism. To this end, we will work with the following generalization of the simplistic model discussed in the previous section:

1. The cell is composed of ribosomes (R), the energy producing machinery (E) and other proteins (O).
2. The cell is growing in a steady state where the cell volume and the cell protein content are growing at the same rate μ.
3. The content of other proteins is proportional to the cell volume.
4. There are no other metabolic demands besides the protein synthesis demand of cell growth.

The third postulate assumes that the other proteins mainly account for structural proteins that are responsible for maintaining the cell shape; for example, the proteins that make up the cytoskeleton. This structural support function indicates that their content should scale with the cell volume.

In this simplified model, cell growth can be schematically represented by the effective biochemical reaction

$$R + E + O \xrightarrow{\text{R,E}} 2(R + E + O) \tag{4.2}$$

where the ribosomes are responsible for the catalysis of protein synthesis and the energy producing machinery satisfies the energy demand of protein synthesis. From postulate (1), we can write the balance of volume fractions

$$\phi = \phi_0 + \phi_R + \phi_E \tag{4.3}$$

where ϕ, ϕ_0, ϕ_R and ϕ_E are the cell volume fractions occupied by the cell biomass, other proteins, ribosomes and the energy production machinery. From postulate (1), we can also write the balance of protein content

$$P = P_0 + n_E \phi_E + n_R \phi_R \tag{4.4}$$

where P is the protein concentration in units of amino acids per unit of cell volume, P_0 is the concentration of other proteins, n_R is the number of amino acids per unit of molar volume of ribosome and n_E is the number of amino acids per unit of molar volume of the energy production machinery.

From postulates (2)–(4), we can write the metabolic balances, of protein synthesis/growth

$$\phi_R h_R = \mu P \tag{4.5}$$

and energy production/consumption

$$h_E \phi_E = e_P \mu P \tag{4.6}$$

where h_R is the ribosomes specific horsepower, h_E is the specific horsepower for energy generation and e_P is energy cost of protein synthesis per unit of amino acid.

Eliminating ϕ_R, ϕ_E and P from Eqs. (4.3)–(4.6) and setting $\phi = \phi_{\max}$, we obtain an equation for the growth rate of model (4.2)

$$\mu_{\max} = \frac{1}{\dfrac{1}{\mu_{\max,\,B}} + \dfrac{1}{\mu_{\max,\,MC}}} \tag{4.7}$$

where

$$\mu_{\max,\,B} = \frac{h_R}{n_R} \frac{1}{1 + c_B} \tag{4.8}$$

$$\mu_{\max,\,MC} = \left(\phi_{\max} - \phi_0\right) \frac{h_R}{P_0} \frac{1}{1 + c_{MC}} \tag{4.9}$$

$$c_B = \frac{e_P(h_R/n_R)}{h_E/n_E} \tag{4.10}$$

$$c_{MC} = \frac{e_P h_R}{h_E} \tag{4.11}$$

where the subscript 'max' acknowledges this is the maximum expectation when all precursors are provided in excess and the cell biomass is at its maximum density.

To understand these equations, I will introduce some definitions about the burden of biosynthesis and space allocation on cell growth.

Biosynthesis cost. I will define the biosynthesis cost – denoted by c_B – as the ratio between the total number of amino acids in the energy producing machinery and the total number of amino acids in ribosomes

$$c_B = \frac{n_E \phi_E}{n_R \phi_R} \tag{4.12}$$

The biosynthetic cost is basically the relative amount of amino acids that cannot be used to make more ribosomes because they are required to make the energy producing machinery that satisfy the energy requirements of cell growth. Taking into account the equation of protein synthesis balance (4.5) and energy balance (4.6) we can crosscheck that the definition of biosynthetic cost (4.12) is equivalent to Eq. (4.10).

Molecular crowding cost. I will define the molecular crowding cost – denoted by c_{MC} – as the ratio between the cell volume fraction of the energy producing machinery and the protein synthesis machinery

$$c_{MC} = \frac{\phi_E}{\phi_R} \tag{4.13}$$

The molecular crowding cost is basically the relative amount of intracellular space that is forbidden to the protein synthesis machinery due to the allocation of the energy producing machinery. Taking into account the equation of protein synthesis balance (4.5) and energy balance (4.4), we can crosscheck that the definition of molecular crowding cost (4.13) is equivalent to Eq. (4.11).

With these cost definitions in mind, we can now understand the equations derived above. The maximum growth rate given by Eq. (4.7) is determined by the contribution of two factors: the biosynthesis and molecular crowding limits. The biosynthesis limit $\mu_{max,B}$, Eq. (4.8), tell us that the maximum growth rate cannot exceed the ribosome autocatalytic limit h_R/n_R, times a correction due to the biosynthesis cost of the energy producing machinery.

In turn, the molecular crowding limit $\mu_{max,MC}$, Eq. (4.9), tell us that the maximum growth rate cannot exceed the rate of protein synthesis when the ribosomes are at maximum crowding $(\phi_{max} - \phi_0)h_R/P_0$, times a correction due to the molecular crowding associated with the energy producing machinery. It should be noted that molecular crowding plays a role because of postulate (3). There are other – nonmetabolic – proteins whose content increases in proportion to the cell volume. This implies that $P_0 > 0$ and therefore $\mu_{max,MC}$, as defined by Eq. (4.9), has a meaningful value compared to $\mu_{max,B}$.

4.3 MAXIMUM GROWTH RATE OF BACTERIA

Now, we can improve the estimate for the maximum growth rate of *E. coli* in a rich medium containing glucose and amino acids in excess. To this end, we will use literature reports of the *E. coli* ribosomes translation rate [69], ribosome amino acid content [74] and volume [70], macromolecular composition [71], protein density [72] and proteome composition [75]. Based on our estimates for eukaryote cells,

Table 4.1 *E. Coli* Metabolic Parameters, Including the Estimated Maximum Growth Rate in Rich Medium (μ_{max})

	Parameter	Value	Units	Reference
E. coli cell	ϕ_{max}	0.40	Volume fraction	[71]
	ρ_P	0.25	kg protein/L of cell	[72]
Ribosome	k_R	22	amino acids/s/ribosome	[69]
	v_R	2680	nm^3/ribosome	[70]
		1613	L/mol of ribosome	
	h_R	46	mol amino acid/h/L ribosome	Section 4.1
	N_R	7336	amino acids/ribosome	[74]
	n_R	4.5	mol amino acid/L ribosome	
Other proteins	ψ_0	0.58	Proteome fraction	[76]
	ϕ_0	0.23	Volume fraction	$\phi_{max}\psi_0$
	P_0	1.3	mol amino acid/L of cell	$\rho_P\psi_0/m_a$
	m_a	109	g/mol	Average
Growth rate	ϕ_{max}-ϕ_0	0.17	Volume fraction	
	$\mu_{max,B}$	11	Doublings/h	Eq. (4.8)
	$\mu_{max,MC}$	6.2	Doublings/h	Eq. (4.9)
	μ_{max}	3.9	Doublings/h	Eq. (4.7)
	T_{min}	0.18	h	$\ln 2/\mu_{max}$

fermentation has a higher horsepower than oxidative phosphorylation (Table 2.1). We will therefore assume that fermentation satisfy the energy demand at high growth rates. The fermentation enzymes volume fraction that is required to sustain a high energy demand is very small. Based on this observation we will assume that the biosynthetic and molecular crowding costs of glycolysis are negligible: $c_B \approx 0$ and $c_{MC} \approx 0$. This approximation allows us to obtain an upper bound on the maximum growth rate of *E. coli* in a rich medium (Table 4.1).

The *E. coli* maximum growth rate is estimated to be $\mu_{max} = 3.7$ doublings/h. This value is higher but no far from the observed maximum growth rate of *E. coli* in rich medium, of about 2 doublings/h [75]. In this calculation we can deconvolute the contribution from the biosynthesis and molecular crowding limits. The molecular crowding limit is smaller than the biosynthetic limit, 5.7 versus 11 doublings/h, indicating that molecular crowding has a stronger influence in determining the maximum growth rate of *E. coli*. However, both limits contribute to the final maximum growth rate estimate of 3.7 doublings/h.

We should notice that, based on Eqs. (6.6)–(6.8), the only relevant parameters to make that maximum growth estimate are quantities characterizing the ribosomes, h_R and n_R; macromolecular composition, ϕ_0 and P_0; and the maximum macromolecular packing density ϕ_{max}. These are the key parameters to estimate the maximum growth rate of bacteria in rich medium.

4.4 MAXIMUM GROWTH RATE OF EUKARYOTES CELLS

To estimate the maximum growth rate of eukaryote cells, we introduce two corrections to the calculations made above. First, we need to take into account that, in addition to the protein synthesis demand for growth, eukaryote cells have a basal rate of protein turnover. For mammalian cells, this is about 0.01 proteins/h [77], which is close to typical growth rates of mammalian cells (1 doubling/8–48 h or 0.04–0.01 doublings/h). Second, eukaryote cells have a high energy demand for cell maintenance even in the absence of cell proliferation. To account for these two factors, we need to rewrite the equation for protein synthesis balance (4.5) and energy balance (4.6) as follows:

$$\phi_R h_R = (k_P + \mu)P \tag{4.14}$$

$$h_E \phi_E = m_E + e_P(k_P + \mu)P \tag{4.15}$$

where k_P is the basal rate of protein turnover rate and m_E is the energy demand of cell maintenance: moles of ATP consumed per unit of cell volume and time. The equations that will follow are the same as (4.3)–(4.6) after performing the following transformations:

$$
\begin{aligned}
\mu &\to k_P + \mu \\
P_0 &\to P_0 + \frac{n_E m_E}{h_E} \\
\phi_0 &\to \phi_0 + \frac{m_E}{h_E} \\
\phi_E &\to \phi_E - \frac{m_E}{h_E}
\end{aligned}
\tag{4.16}
$$

Their interpretation is straightforward. We need to replace the demand for growth by the demand for growth and protein turnover. To the basal protein content of cells P_0, we need to add the protein content associated with the energy production machinery that satisfies the maintenance energy demand. Finally, to the basal volume fraction ϕ_0, we need to add the volume fraction occupied by the energy production machinery that satisfies the maintenance energy demand. These are basically the different contributions of growth independent factors to the cell metabolism and macromolecular composition. After performing these transformations, Eqs. (4.7)–(4.9) are replaced by

$$\mu_{\text{max}} = \frac{1}{\dfrac{1}{\mu_{\text{max}, B}} + \dfrac{1}{\mu_{\text{max}, \text{MC}}}} - k_P \tag{4.17}$$

where

$$\mu_{\text{max}, B} = \frac{h_R}{n_R} \frac{1}{1 + c_B} \tag{4.18}$$

$$\mu_{\text{max}, \text{MC}} = \left(\phi_{\text{max}} - \phi_0 - \frac{m_E}{h_E} \right) \frac{h_R}{P_0 + \dfrac{n_E m_E}{h_E}} \frac{1}{1 + c_{\text{MC}}} \tag{4.19}$$

From the comparison of Eqs. (4.8) and (4.18), we note that the biosynthesis limiting growth rate $\mu_{\text{max}, B}$ remains the same. This was expected given that the biosynthesis limit is basically the autocatalytic limit due to the fact that the biosynthesis machinery synthesizes itself. It does not depend on any factor that is not related to the synthesis of the biosynthesis machinery. In contrast, the molecular crowding limit has noticeable changed from Eqs. (4.9)–(4.19). The volume fraction available to the biosynthesis machinery has been reduced after the allocation of the energy producing machinery to satisfy the energy maintenance demand. The amount of protein that needs to be synthesized independently of biosynthesis has increased, after the requirement of additional protein to make the energy producing machinery that satisfies the energy maintenance demand. These two factors will ultimately lead to a dramatic reduction in the maximum growth rate of eukaryote cells.

First, we estimate the maximum growth rate of mammalian cells. To this end, we will use literature reports of the ribosomes translation rate [78], ribosome amino acid content [79] and volume [80], energy cost of translation [81], macromolecular composition [71], protein density [82], proteome composition [75], basal rate of protein turnover [77] and energy demand for cell maintenance [83]. For mammalian cells, we have estimated the specific horsepower of oxidative phosphorylation and fermentation for energy generation (Table 2.1). This will allow us to make a full calculation of the maximum growth rate that can be sustained by either pathway (Table 4.2). Assuming an oxidative

Table 4.2 Mammalian Cell Parameter Estimates, Including the Estimated Maximum Growth Rate in Rich Medium Using Oxidative Phosphorylation or Fermentation

	Parameter	Value	Units	Reference
Mammalian cell	ϕ_{max}	0.40	Volume fraction	
	ρ_P	0.14	kg protein/L of cell	[82]
Ribosome	k_R	5.6	Amino acids/s/ribosome	[78]
	v_R	4000	$(nm)^3$	[80]
		2408	L/mol	
	h_R	8.4	mol amino acid/h/L ribosome	
	N_R	11,590	Amino acid/ribosome	[79]
	n_R	4.8	mol amino acid/L ribosome	
	$\mu_{max,R}$	2	Doublings/h	
	e_P	4.6	ATP/amino acid	[81]
Other proteins	ψ_0	0.85	Proteome fraction	[76]
	ϕ_0	0.34	Volume fraction	$\phi_{max}\psi_0$
	P_0	1.9	mol amino acid/L of cell	$\rho_P\psi_0/m_a$
	m_a	109	g/mol	Average
Maintenance	k_P	0.01	Protein turnover/h	[77]
	m_E	17	pmol ATP/cell/day	[83]
		0.21	mol ATP/L/h	
OxPhos	h_M	10	mol ATP/h/L of mitochondria	Table 2.1
	v_s	3	L/kg protein	
	n_M	3.6	mol amino acids/L	
	$c_{B,O}$	2.8		Eq. (4.12)
	$c_{MC,O}$	3.9		Eq. (4.11)
	$\phi_{ME,O}$	0.021		
	$P_{ME,O}$	0.074		
	$\mu_{max,B,O}$	0.45	Doublings/h	Eq. (4.18)
	$\mu_{max,MC,O}$	0.06	Doublings/h	Eq. (4.19)
	$\mu_{max,O}$	0.04	Doublings/h	Eq. (4.17)
	$T_{min,O}$	17	h	
Fermentation	h_F	91	ATP/h/L of mitochondria	Table 2.1
	v_s	0.79	L/kg protein	
	n_F	11.6	mol amino acids/L	
	$c_{B,F}$	1.0		Eq. (4.12)
	$c_{MC,F}$	0.4		Eq. (4.11)
	$\phi_{ME,F}$	0.002	Volume fraction	
	$P_{ME,F}$	0.027	mol amino acid/L of cell	
	$\mu_{max,B,F}$	0.86	Doublings/h	Eq. (4.18)
	$\mu_{max,MC,F}$	0.30	Doublings/h	Eq. (4.19)
	$\mu_{max,F}$	0.13	Doublings/h	Eq. (4.17)
	T_{min}	3	h	

phosphorylation specific horsepower of 10 ATP/h/L of mitochondria, which is about the median of the values reported in Table 2.1, we obtain a minimum doubling time of 17 h. Using fermentation mammalian cells can proliferate at a faster rate, with a minimum doubling time of 3 h. This is of course an approximate value. We have not taken into account the growth requirement of lipids and nucleotide synthesis. However, given that protein is the major component of the cell biomass (60% in mammalian cells [68]), this estimate has given us a reasonable order of magnitude.

The predicted minimum doubling time of mammalian cells (3 h) is 10 times higher than the *E. coli* estimate (0.25 h). There are different factors contributing to this difference. By comparison of Tables 4.1 and 4.2, we notice that mammalian ribosomes are larger and translate at a lower rate. These changes probably occurred during evolution to improve the fidelity of translation, but they came at the expense of a sixfold decrease in the ribosome horsepower. There is also an increase in the basal energy requirements for cell maintenance. Here again, this probably occurred during evolution to protect the integrity of the mammalian cell and its genome.

Based on the estimates in Table 4.2, in the range of doubling times between 3 and 17 h mammalian cells require fermentation to sustain the energy demands of cell growth. Mitochondria do not have sufficient horsepower to sustain growth with a doubling time below 17 h. Just to put this numbers into perspective, during an immune response to a viral infection, the population of T-cells expands at a very high rate, reaching doubling times as low as 8 h [84]. Since the T-cells doubling time falls within the window where fermentation is obligatory, we predict that during an immune response T-cells ferment glucose to lactate. And indeed they do [85,86].

Similarly, we can estimate the maximum growth rate of yeast cells. Yeast cells are similar in structure and composition to mammalian cells, and therefore, in most cases, we will use parameter values reported for mammalian cells (Table 4.2). The few exceptions we make are the proteomic fraction of nonmetabolic proteins, which is lower in yeast cells (37% [87]) than in mammalian cells (85% [76]). And the mitochondria horsepower, which is higher in yeast cells (17 mol ATP/L/h, Table 2.1) than the median value for healthy mammalian cells (10 mol ATP/L/h, Table 2.1). The outcome of these estimates is reported in Table 4.3. Assuming pure oxidative phosphorylation metabolism we

Table 4.3 Yeast Cell Parameter Estimates, Including the Estimated Maximum Growth Rate in Rich Medium Using Oxidative Phosphorylation or Fermentation

	Parameter	Value	Units	Reference
Mammalian cell	ϕ_{max}	0.40	Volume fraction	
	ρ_P	0.14	kg protein/L of cell	[82]
Ribosome	k_R	5.6	Amino acid/s/ribosome	[78]
	v_R	4000	$(nm)^3$	[80]
		2408	L/mol	
	h_R	8.4	mol amino acid/h/L ribosome	
	N_R	11,590	Amino acid/ribosome	[79]
	n_R	4.8	mol amino acid/L ribosome	
	$\mu_{max,R}$	2	Doublings/h	
	e_P	4.6	ATP/amino acid	[81]
Other proteins	ψ_0	0.37	Proteome fraction	[87]
	ϕ_0	0.15	Volume fraction	$\phi_{max}\psi_0$
	P_0	0.5	mol amino acid/L of cell	$\rho_P\psi_0/m_a$
	m_a	109	g/mol	Average
Maintenance	k_P	0.01	Protein turnover/h	[77]
	m_E	1	mmol ATP/g dry weight/h	[88]
		0.3	mol ATP/L/h	$\times$ 0.3 kg/L
OxPhos	h_M	17	mol ATP/h/L of mitochondria	Table 2.1, Yeast
	v_s	3	L/kg protein	
	n_M	3.6	mol amino acid/L	
	$c_{B,O}$	1.6		Eq. (4.12)
	$c_{MC,O}$	2.2		Eq. (4.11)
	$\phi_{ME,O}$	0.012		
	$P_{ME,O}$	0.053		
	$\mu_{max,B,O}$	0.66	Doublings/h	Eq. (4.18)
	$\mu_{max,MC,O}$	1.21	Doublings/h	Eq. (4.19)
	$\mu_{max,O}$	0.42	Doublings/h	Eq. (4.17)
	$T_{min,O}$	2	h	
Fermentation	h_F	91	ATP/h/L of mitochondria	Table 2.1
	v_s	0.79	L/kg protein	
	n_F	11.6	mol amino acid/L	
	$c_{B,F}$	1.0		Eq. (4.12)
	$c_{MC,F}$	0.4		Eq. (4.11)
	$\phi_{ME,F}$	0.002	Volume fraction	
	$P_{ME,F}$	0.027	mol amino acid/L of cell	
	$\mu_{max,B,F}$	0.86	Doublings/h	Eq. (4.18)
	$\mu_{max,MC,F}$	2.9	Doublings/h	Eq. (4.19)
	$\mu_{max,F}$	0.66	Doublings/h	Eq. (4.17)
	T_{min}	1	h	

obtain a maximum growth rate of about 0.4 doublings/h, corresponding to a minimum doubling time of 2 h. Using fermentation yeast cells can proliferate at a faster rate of 0.66 doublings/h, corresponding to a minimum doubling time of 1 h. It is interesting to notice that, as a difference to *E. coli* (Table 4.1) and mammalian cells (Table 4.2), in yeast cells the maximum growth rate derived from the biosynthetic cost alone is lower than that derived from the molecular crowding cost alone (Table 4.3, $\mu_{max,B,O}$ vs $\mu_{max,MC,O}$ and $\mu_{max,B,F}$ vs $\mu_{max,MC,F}$). This is because the reported proteomic fraction of nonmetabolic proteins is lowest in yeast cells ($\psi_0 = 0.37$, Table 4.3) compared to *E. coli* ($\psi_0 = 0.58$, Table 4.3) and mammalian cells ($\psi_0 = 0.85$, Table 4.3). As we mentioned earlier, the impact of the molecular crowding cost is only relevant because of the existence of this component of nonmetabolic proteins. The lower fraction of nonmetabolic proteins in yeast cells relative to mammalians cells is also the factor allowing yeast cells to proliferate faster.

4.5 THE CHEMOSTAT EXPERIMENT

The *chemostat* is an experimental apparatus where the chemical environment can be maintained static and nutrient availability can be controlled by the experimenter. This is achieved by culturing cells in a vessel subject to continuous supply of nutrients and continuous overflow of content exceeding the vessel volume (a glass under a dripping tap is good analogy). The chemical environment in the chemostat vessel is determined by the chemical composition of the feeding medium, the dilution rate (rate of liquid volume addition relative to the vessel volume) and the type of cells cultured. After some transient period, the proliferation rate of the cells in the vessel will match the dilution rate [89], provided that the cells remain in suspension and the dilution rate does not exceed the maximum proliferation rate.

Cells need sources of carbon, nitrogen and other elements to proliferate. The relative abundance of these sources in the feeding medium determines which nutrient is the limiting factor for growth. For every chemostat experiment we need to specify the limiting element: carbon, nitrogen, etc. It is also common to name the experiment after the specific source of limiting element. For example, in a glucose-limited experiment, carbon is the limiting element and glucose is the carbon source. We also need to specify whether the experiment was conducted

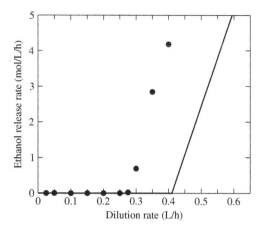

Figure 4.1 Ethanol excretion rate by yeast cells. The symbols represent measurements in units of mol ethanol/g dry weight/h [90], multiplied by the typical biomass density of a cell 0.3 kg/L. The line represents the theoretical prediction from Eqs. (4.24) and (4.25).

in the presence (aerobic) or absence (anaerobic) of oxygen. The chemostat is the experimental apparatus of choice to investigate the metabolism of cells that can grow in suspension at different proliferation rates. Since proliferation rate is in this context a measure of metabolic rate, we can use the chemostat to investigate the differential utilization of oxidative phosphorylation and fermentation at different metabolic rates.

Fig. 4.1 shows the typical output of a chemostat experiment for a glucose-limited culture of yeast cells under aerobic conditions, based on data reported in reference [90]. At low dilution rates there is no fermentation. I will call this scenario the Pasteur phase: 'oxidative phosphorylation represses glycolysis'. However, after a threshold dilution rate fermentation kicks in and its rate increases linearly with increasing the dilution rate. I will call this scenario the Brown−Warburg−Crabtree (BWC) phase, where aerobic fermentation is manifested.

The transition from pure oxidative phosphorylation in the Pasteur phase to mixed oxidative phosphorylation with fermentation in the BWC phase is a consequence of two simple facts pointed in the previous section. Oxidative phosphorylation has a higher yield of ATP per molecule of glucose than fermentation but fermentation can sustain a faster proliferation rate. When the proliferation rate exceeds the maximum growth rate allowed by oxidative phosphorylation ($\mu_{\text{max},o}$) fermentation becomes obligatory.

To model the concomitant occurrence of aerobic and fermentation metabolism we extend the cell growth model considered above after dividing the energy production by the corresponding components associated to oxidative phosphorylation and fermentation. This result in the updated volume fractions balance

$$\phi = \phi_0 + \phi_R + \phi_F + \phi_O \qquad (4.20)$$

where ϕ_F and ϕ_O are the cell volume fractions occupied by the fermentation and oxidative phosphorylation machinery, respectively. From postulate (1), we can also write the balance of protein content

$$P = P_0 + n_F\phi_F + n_O\phi_O + n_R\phi_R \qquad (4.21)$$

where n_F is the number of amino acids per unit of molar volume of fermentation enzymes and n_O is the number of amino acids per unit of molar volume of the oxidative phosphorylation machinery.

From postulates (2)–(4), we can write the metabolic balances, of protein synthesis/growth

$$\phi_R h_R = \mu P \qquad (4.22)$$

and energy production/consumption

$$h_F\phi_F + h_O\phi_O = e_P\mu P \qquad (4.23)$$

where h_F and h_O are the horsepower for energy generation by fermentation and oxidative phosphorylation and e_P is energy cost of protein synthesis per unit of amino acid.

Using Eqs. (4.20)–(4.23), we can determine the combinations of aerobic and fermentation metabolism that can sustain the energy demand of cell growth as a function of the growth rate (Fig. 4.2). The solutions go from all energy being generated by fermentation (assuming that there is enough sugar to do so) to the solution with minimum fermentation. We can estimate the minimum fermentation rate that is required to sustain growth $- r_{F,min}(\mu) -$ by interpolating between the lack of absolute fermentation requirement at $\mu_{max,O}$ and the absolute requirement at $\mu_{max,F}$

$$r_{F,\,min}(\mu) = \begin{cases} 0 & \mu \geq \mu_{max,\,O} \\ \dfrac{\mu - \mu_{max,\,O}}{\mu_{max,\,F} - \mu_{max,\,O}} r_{F,\,max} & \mu_{max,\,O} < \mu \leq \mu_{max,\,F} \end{cases} \qquad (4.24)$$

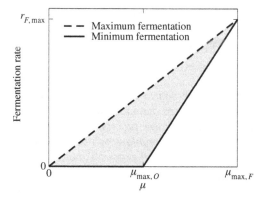

Figure 4.2 Combinations of aerobic and fermentation metabolism satisfying the energy demand at different growth rates.

where $r_{F,\max}$ is the maximum fermentation rate at the maximum growth rate sustained by fermentation. $r_{F,\max}$ is obtained from Eqs. (4.20)–(4.23) setting $\phi_O = 0$ and $\mu = \mu_{\max,F}$ resulting in

$$r_{F,\max} = \frac{e_P \mu_{\max,F}\left(P_0 + n_R\left(\phi_{\max} - \phi_0\right)\right)}{1 + e_P \mu_{\max,F}(n_R - n_F)/h_F} \qquad (4.25)$$

Using the parameter estimates for yeast cells (Table 4.3) and Eqs. (4.24)–(4.25), we obtain the theoretical line depicted in Fig. 4.2. It clearly captures the qualitative behaviour of the experimental data. No fermentation below the threshold growth rate and a linear increase of fermentation above the threshold. The onset of overflow metabolism is basically the maximum growth rate that can be sustained by oxidative phosphorylation ($\mu_{\max,O}$). The theoretical estimate is slightly higher than what observed. Yet, it is quite a good estimate given that we only took into consideration the protein requirements of cell growth. Addition of other cell components, particularly lipids, will certainly reduce the theoretical estimate of the maximum growth rate that can be sustained by oxidative phosphorylation.

4.6 BIOSYNTHETIC, MOLECULAR CROWDING AND PROTEOMIC COST

At this point, I would like to bring up a discussion about the relative contribution of biosynthetic cost and molecular crowding to overflow metabolism. For reasons beyond my understanding, there is some resistance to accept molecular crowding as a key factor behind overflow metabolism. In contrast, biosynthetic cost has been the hypothesis of choice for most of the research community. The oldest reference I have found dates to 1999, in a book about the principles of wine making by Boulton et al. [91]:

> *Why would cells evolve preferring to ferment even in the presence of oxygen when more ATP can be produced during respiration per glucose molecule consumed? Respiration is an expensive process, requiring more in the way of enzymatic machinery and mitochondria which can be dispensed with during fermentation. Thus, while respiration produces more energy, more energy is consumed in maintaining respiratory capacity. If sugar is not limiting, fermentation allows a fast growth rate and the same total amount of ATP can be produced. In grape juice, sugar is never the limiting nutrient, meaning that carbon and energy are plentiful.*

The statement that more energy is consumed in maintaining respiratory capacity is the implicit reference to biosynthetic cost. I have heard similar arguments in scientific publications, meetings and personal discussions.

However, can the biosynthetic cost alone explain overflow metabolism? To answer that question, we can go back the calculations made in previous sections and remove the contribution of molecular crowding. That is equivalent to set $\mu_{max} = \mu_{max,B}$. Within this assumption, the maximum growth rate that can be sustained by a mammalian cell without fermentation is 0.04 doublings/h ($\mu_{max,B,O}$, Table 4.2), which translates to a doubling time of about 2 h. This value cannot explain why T-cells proliferating at a lower rate, with doubling times around 8 h, manifest high rates of fermentation. The biosynthetic cost alone cannot explain overflow metabolism as observed in proliferating cells. We need to consider the contribution of molecular crowding. Eqs. (4.7) and (4.17) clearly demonstrate that maximum growth rates are determined by both the biosynthesis and molecular crowding limits on growth rate.

In 2015, the Hwa's group pushed forward the hypothesis that overflow metabolism is the consequence of a higher proteomic cost of energy biogenesis by respiration [92]:

> *Here we study metabolic overflow in Escherichia coli, and show that it is a global physiological response used to cope with changing proteomic demands of energy biogenesis and biomass synthesis under different growth conditions. A simple model of proteomic resource allocation can*

quantitatively account for all of the observed behaviours, and accurately predict responses to new perturbations. The key hypothesis of the model, that the proteome cost of energy biogenesis by respiration exceeds that by fermentation, is quantitatively confirmed by direct measurement of protein abundances via quantitative mass spectrometry.

However, what does proteomic cost really means? According to the definition given by Hwa's group, proteomic cost is the relative contribution of a protein mass or the protein mass associated with a compartment or pathway to the total protein mass in the cell. While there is nothing wrong about the definition of proteomic cost, we should bear in mind that it is one of many types of cost we could define.

We have already defined the biosynthetic cost, the relative number of amino acids in a protein, organelle or pathway. The biosynthetic cost appears naturally when we investigated cell growth. It takes into account that when cells proliferate there is a cost involved in duplicating every cell component. We have also dealt with the molecular crowding cost, the relative volume fraction occupied by a protein, organelle or pathway. It takes into account that the cell components have a finite size limiting how high they can be concentrated.

In contrast, I cannot find a natural explanation for the proteomic cost. The mass of a protein would be relevant if inertia or gravity would play a role. However, in the highly viscous intracellular milieu, the influence of inertia or gravity on the dynamics of proteins and organelles is negligible. Therefore, in matters of cell metabolism, the only relevance of proteomic cost would be as a surrogate of a *bona fide* cost. And indeed, molecular mass is a good approximation to both the biosynthetic and molecular crowding costs. For example, the amino-acid content and volume of a globular protein are proportional to the protein mass.

In the body of work from Hwa's group, the proteomic cost is basically a surrogate of biosynthetic cost. Since the biosynthetic cost is a natural concept the work of Hwa should actually be read as: 'the *biosynthetic* cost of energy biogenesis by respiration exceeds that by fermentation', where biosynthetic has replaced proteome in the original source [92]. Of course, it seems we are getting tangled into the semantics. The truth of the matter is that biosynthetic cost is a more general term and it predates Hwa's group work. We could extend the calculations made in this chapter to accommodate the RNA composition of ribosomes. We will obtain modified equations to the biosynthetic cost in Eq. (4.12). However, the term proteomic cost will not apply anymore because it will include the cost of RNA synthesis. Yet, the term biosynthetic cost will still be valid. It is also worth noticing that the molecular crowding cost is neglected altogether in the calculations reported by Hwa's group [92]. A technical discussion of how and where the molecular crowding cost becomes relevant in those calculations can be found here [93].

4.7 GENOME-SCALE MODELS

The calculations made above can be extended to include all requirements of cell growth besides protein. They can also be extended to deal with less rich nutrient conditions, where cells are forced to synthesize amino acids and other precursors to survive. While this can be done with pen and paper, it is more efficiently done using computational models. These models allow us to handle a large number of mass conservation constraints, and they facilitate the model update upon acquisition of new knowledge.

The development of full-scale metabolic models of cells has been aided by the sequencing of different species genomes. Using the newly derived genome sequences together with cross-species genome alignment and annotations of gene function one can build genome-scale models of cell metabolism [94]. These models contain all the putative enzymes that are encoded on the organism under consideration. The genome-scale annotations are just the backbone of metabolic models. The network of biochemical reactions is quite redundant when it comes to fulfil the synthesis of a specific compound from a specific substrate. Without any additional information, the number of possible ways to duplicate the content of the cell increases exponentially with the size of the number of biochemical reactions. Additional constraints must be added to narrow down the space of possible solutions that are consistent with known physicochemical constraints [95]. This has led to whole phylogeny of modelling approaches to tackle the redundancy of cell metabolism and improve on model predictions [96,97]. In this phylogeny, there is one − tiny − branch corresponding to models accounting for the volume allocation (molecular crowding) constraint. These are the type of models described below.

In the context of *E. coli* metabolism, there have been gradual improvements on the estimation of maximum growth rates and the threshold to acetate excretion. In the first implementation of the volume allocation constraint, the effective turnover numbers relating metabolic fluxes to enzyme concentrations were sampled from an empirical distribution [98]. This approach lead to good estimates of the maximum growth rate *E. coli* cells in different nutrients. These estimates were later improved by Adadi et al. [99], after introducing better annotations of the turnover number of each enzyme. To my knowledge, the latter are the most precise estimates of *E. coli* maximum growth rates in different nutrients.

Genome-scale models have also been applied to explain the switch from pure oxidative phosphorylation to mixed oxidative phosphorylation and fermentation by *E. coli* cells grown in glucose-limited chemostats [100,101]. These models included biosynthetic costs implicitly and the volume allocation constraint explicitly. Yet, at the point of making conclusions about which factor was responsible for the metabolic switch, Vazquez et al. [100] emphasized the volume allocation constraint and Molenaar et al. [101] emphasized the biosynthetic cost. As demonstrated here, both factors are relevant to determine the maximum growth rate using oxidative phosphorylation, which determines the threshold growth rate to overflow metabolism. Years later Basan et al. [92] claimed that they have explained overflow metabolism introducing the concept of proteomics cost. But, as mentioned in Section 4.7, their work is just a reformulation of the biosynthetic cost with an implicit volume allocation constraint given by a fixed protein density (see Ref. [93] for further discussion).

Similar work has been carried on in the context of yeast metabolism [87]. Nilsson and Nielsen implemented and analysed a computational model of yeast metabolism with explicit annotation of the turnover numbers associated with each enzyme. Their model provides a bettwer quantitative agreement for the switch to aerobic fermentation by yeast cells than obtained from our simplified calculations (Fig. 4.1).

In the context of human metabolism, Vazquez et al. [37] demonstrated that energy balance together with the volume allocation constraint can explain the manifestation of overflow metabolism in cancer cells. Later on, Shlomi et al. [102] reported the first genome-scale model accounting for protein allocations constraints, which was followed by our own implementation [103]. Interestingly, why both reports recapitulated the observed increase on glutamine consumption by highly proliferating mammalian cells, no explanation was provided of why glutamine consumption increased. Years later, armed with pen and paper, I was able to show that glutamine consumption increases because the mitochondria have a limited capacity to turn over the NADH generated during aerobic biosynthesis from glucose [38]. I make an emphasis on the use of pen and paper in an attempt to convince you that, even in the era of computers, simple calculations are still a good way to get at the root of a problem.

It is worth mentioning that Adadi et al. [99], Nilsson and Nielsen [87] and Shlomi et al. [102] worked with proteomic cost rather than directly with the volume allocation or molecular crowding constraint. This is becoming a trend in the metabolic modelling field, and it is motivated by the recent advances in mass spectrometry for the quantification or protein abundance at the genome scale. From the operational point of view, it does not make much of a difference to work from scratch and use the volume allocation constraint, or to start at one step below and use the protein allocation constraint under the assumption that the protein density is constant. On the other hand, from the conceptual point of view, we should bear in mind that the observed protein density of cells is a direct consequence of molecular crowding.

The problem of heterogeneous cell populations interacting via the release and consumption of metabolites has also been addressed in recent years [104–106]. Using back of the envelope calculations or genome-scale models they have unanimously shown that the primary cells (e.g., neurons or cancer cells) benefit from a second population of supporting cells assimilating fermentation products.

Yet, I feel the jump in quality will come when full kinetic models of each biochemical reaction are included into the genome-scale formulation. The major challenge is the estimation and sampling of missing kinetic parameters. Workflows to address this problem have been reported and applied to genome-scale models of different organisms [107–109]. An iterative reconstruction approach has been also proposed for the reconstruction of the metabolic network, reaction kinetic laws and kinetic parameters [110]. The good news is that the volume allocation constraint restricts the space of metabolic models with kinetics to elementary flux modes satisfying the metabolic objective [111,112]. *Elementary flux modes* were defined as minimal metabolic flux distributions that are both stoichiometrically and thermodynamically feasible [113]. The volume allocation constraint forces cell metabolism into elementary flux modes. Whether this represented a selective advantage for the evolution of molecular crowding is an open question.

CHAPTER 5

Overflow Metabolism in Human Disease

Everything in excess is opposed by nature.

Hippocrates

5.1 INTRODUCTION

In the first chapter, we alluded to the occurrence of overflow metabolism in cancer and immune responses. Warburg's early observations on aerobic fermentation highlighted metabolic differences between cancer cells and normal tissues [8]. In turn, Crabtree demonstrated that normal tissues infected with viruses exhibited aerobic fermentation [9]. Aerobic fermentation may happen for different reasons in those contexts. In the immune response to infections, the immune cells manifest a dramatic increase in their proliferation rate, while maintaining functional mitochondria. Cancer cells do not proliferate as fast as activated immune cells, but cancer cells mitochondria have lower horsepower. In either case, the outcome is the same, an obligatory increase in fermentation to sustain the energy demand of cell metabolism. This increase in fermentation is similar to the switch to fermentation metabolism when running beyond the lactate threshold speed. The only difference is the tissue where fermentation has increased. In the following, we adapt the fermentation metabolism model of muscle physiology to other fermenting tissues.

5.2 BASIC MODEL OF FERMENTATION METABOLISM

In fermentation metabolism peripheral tissues take care of turning over lactate via the Cori cycle [1]

$$
\begin{aligned}
\text{glucose} + 2\,\text{ADP} &\rightarrow \text{lactate} + 2\,\text{ATP} &&\text{Fermenting tissue} \\
\text{lactate} + 3\,\text{ATP} &\rightarrow \text{glucose} + 3\,\text{ADP} &&\text{Peripheral tissue}
\end{aligned}
$$

where lactate is turned over into glucose by gluconeogenesis in peripheral tissue. We are going to base our general model of fermentation metabolism on the following postulates:

Overflow Metabolism. DOI: https://doi.org/10.1016/B978-0-12-812208-2.00005-6

1. The fermenting tissue generates lactate at a rate

$$r_F = v_F f \tag{5.1}$$

where v_F is the specific volume of fermenting tissue (litres per kg of body weight) and f is the speed of fermentation at the fermenting tissue (mol lactate/litre of tissue/hour).

2. The lactate production and consumption are balanced through the Cori cycle, whereby the energy demand of gluconeogenesis from lactate ($3/2 r_F$) is sustained by aerobic metabolism at peripheral tissues

$$\frac{3}{2} r_F = v_P \phi_{PM} h_M = \varepsilon s^* \tag{5.2}$$

where s^* is the equivalent running speed that would amount to the same rate of fermentation. From Eqs. (5.1) and (5.2), we obtain the equivalent running speed

$$s^* = \frac{3}{2\varepsilon} r_F = \frac{3}{2\varepsilon} v_F f \tag{5.3}$$

The equivalent running speed is a theoretical construction to achieve a better intuition about the metabolic burden caused by the fermenting tissue under investigation. Using this quantity we can apply what we learned in Chapter 3 about fermentation metabolism in muscle physiology to any fermenting tissue.

5.3 IMMUNE RESPONSES

Lactate blood levels are found elevated during immune responses to infections. The extreme case is sepsis, a life-threatening condition caused by an acute immune response triggered by an infection [114]. The increase in blood lactate levels in sepsis patients is so high that it has been considered as a diagnostic tool [114]. As it is generally assumed, lack of oxygenation is hypothesized to be the cause of increased fermentation. Yet, we have demonstrated that, even in conditions with no oxygen limitation, fast cell proliferation demands fermentation. The proliferation rate of immune cells increases dramatically resulting in doubling times below 17 h. Such fast proliferation cannot be sustained by pure aerobic metabolism and some degree of fermentation is obligatory (Chapter 4). Therefore, in the course of an immune response, there will be some finite rate of fermentation f per unit of active immune cell volume. That multiplied by the specific volume of activated immune cells will give us the total rate of fermentation associated with the immune response. The overall specific volume accounted by proliferating immune cells is given by

$$v_F = \frac{N V_{\text{cell}}}{W} \tag{5.4}$$

where N is the number of active immune cells, V_{cell} is the typical cell volume and W is the body weight. Using typical values, we can estimate the effective running speed that would account for the same rate of fermentation associated with an immune response. For example, a population of 10^{12} active immune cells fermenting at a rate of 1 mol lactate/L of cell/h will incur into a metabolic burden comparable to running at a speed of 1 km/h (Table 5.1). From that number, we can extrapolate the effective metabolic speed for larger populations of fermenting cells. When the number of active immune cells exceeds 10^{13}, we can get into a context where the equivalent running speed exceeds the lactate threshold speed (Fig. 5.1). Note that 10^{12} cells amount to a specific volume of 0.01 L/kg and 10^{13} to 0.1 L/kg (assuming cell volume of 1 pL, buoyant cell density 1 kg/L and 70 kg of body weight). The latter number is on the range of the lower body skeletal muscle specific volume: 0.21 [63]. In such an event, blood lactate levels are predicted to increase following Eq. (3.25). In turn, the clinical observation of increased lactate levels in patients with acute immune responses is an indication that the number of activated immune cells is of the order of 10^{13} cells.

Table 5.1 Effective Running Speed Associated with an Immune Response Involving the Proliferation of 10^{12} Cells

Parameter	Value	Units	Reference
Lactate release	1	mol/L of cell/h	Typical value [82]
Number of cells	1	10^{12}	Assumed
Cell volume	1	pL	Typical value [82]
Body weight	70	kg	Assumed
Specific volume	0.01	L/kg of body weight	Calculated
r_F	14	mmol/h/kg	Calculated
ε	22	mmol/km/kg	Chapter 3
s^*	1	km/h	Eq. (5.3)

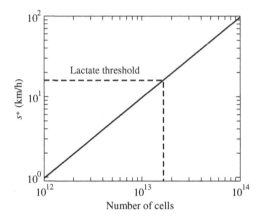

Figure 5.1 Effective running speed associated with a cell population fermenting at a rate of 1 mol/L of cell/h. The horizontal line marks the lactate threshold speed.

5.4 CANCER

The threshold for overflow metabolism in mammalian cells — $\mu_{max,O}$ — is a monotonic increasing function of the mitochondria horsepower for energy generation. The parameter estimates reported in Table 4.2 are based on a median mitochondria horsepower of 10 mol ATP/h/L of mitochondria. However, cancer cells mitochondria exhibit a lower horsepower of about 3 ATP/h/L of mitochondria (Table 2.1). With this low horsepower, the only way cancer cells could satisfy their energy requirements is by increasing the mitochondria content. But as we have been discussing, there is a limit to how high the mitochondria content can be. With that low horsepower, the amount of mitochondria needed to sustain just the energy demand of cell maintenance (m_E, Table 4.2) is 7.5% of the cell volume. That together with the 34% occupied by non-metabolic proteins (ϕ_0, Table 4.2) results in a 41% of the cell volume. However, the macromolecular fraction of the cell does not exceed 40% of the cell volume. Therefore, cancer cells cannot or can barely support the energy demand of cell maintenance using oxidative phosphorylation. Cancer cells should manifest fermentation even when they are not proliferating. And indeed, the energy demand of cancer cells is above what can be supported by oxidative phosphorylation [37].

The key factors leading to overflow metabolism in cancer cells are the volume allocation constraint together with the reduction in the mitochondria horsepower relative to that observed for muscle cells. The latter resonates with Warburg's original hypothesis: cancer cells exhibit high rates of fermentation because they have less efficient mito-chondria. The correction we introduce is that the mitochondria does not need to be fully dysfunctional. A decrease in the mitochondria horsepower – from 10 down to 3 mol ATP/h/L mitochondria – is sufficient to force fermentation as an obligatory phenotype.

Reports of cancer patients with high tumour burden demonstrate that blood lactate levels reach values in the range between 4 and 8 mmol/L [115], as high as observed above the lactate threshold in muscle physiology (Fig. 3.3). A case study of a patient with lymphoma revealed spikes of even higher values of 18 mmol lactate/L during the time course of the disease [116]. In those instances, we would predict cancer-related fermentation metabolism with equivalent running speeds exceeding the lactate threshold (Fig. 5.1).

At this point, we should remember all the physiological changes that take place when we are running at a speed exceeding the lactate threshold. Increased blood circulation, increased oxygen consumption and increased peripheral muscle waste to sustain the Cori cycle. All these physiological changes are expected to happen in cancer patients where the equivalent running speed exceeds the patients lactate threshold. This could explain why a significant per cent of cancer patients suffers from the wasting syndrome known as Cachexia [117].

5.5 NEURODEGENERATION

The formation of intracellular aggregates is a common aetiology of several degenerative diseases [118–120]. The extracellular and intracellular accumulation of the aggregates of β-amyloid protein has been observed in certain brain areas of Alzheimer's patients [120], α-synuclein in Parkinson's patients [118, 120], and huntington in Huntington's patients [121, 122]. Beyond neurodegenerative diseases, the intracellular accumulation of β-amyloid protein aggregates has been hypothesized as a cause of inclusion-body myositis [123, 124], a muscular degenerative disease that starts after age 50 years. The connection between accumulation of intracellular aggregates and degeneration is supported by in vitro studies reporting a negative correlation between the accumulation of intracellular aggregates and cell survival [125, 126].

Similar to the Warburg hypothesis for cancer, mitochondrial defects and oxidative stress has been pointed as the major mechanistic links between the accumulation of intracellular aggregates and cell death [126–128]. Impaired mitochondrial biogenesis contributes to mitochondrial dysfunction in in vitro models of Alzheimer's disease [129]. Mouse models of familial Alzheimer's disease have provided evidence indicating that defects in mitochondria trafficking and integrity precede the onset of neurological phenotype [130]. They have also shown that the distribution of mitochondria is disrupted by the formation of protein aggregates in neurons [131]. Taking together this evidence points to reduced mitochondrial activity as a major factor in Alzheimer's disease.

Brain metabolism has been studied extensively using different experimental techniques. Time resolved experiments have shown that the early neuronal activity (up to about 10 s) is supported by oxidative phosphorylation of lactate from an extracellular pool [132–134]. This initial phase is followed by the activation of the astrocyte-neuron lactate shuttle, whereby astrocytic glycolysis generates lactate that is then utilized by the neurons [132–134]. This evidence indicates that normal neurons undergoing minor neural activity satisfy their energy demands through the oxidative phosphorylation of lactate. However, high lactate levels have been observed in brain regions of Huntington's disease patients compared to normal controls [135, 136], suggesting a shift from a net consumption to a net lactate production by the astrocyte-neuron system.

The progressive increase of intracellular protein aggregates may also constraint the intracellular space available to metabolic enzymes and mitochondria [137]. The cell has a high density of macromolecules occupying about 30%–40% of the intracellular space [68, 71]. An increase of the macromolecular density beyond this value hinders dramatically the diffusion of metabolites and macromolecules, resulting in an exponential reduction of the rate of diffusion-limited reactions [138, 139]. The cell molecular machinery thus operates under a macromolecular crowding constraint, whereby the concentration of macromolecules should not exceed the limiting value of about 40%.

In previous chapters, we have shown that, because mitochondria has a lower horsepower than glycolysis, there is a metabolic switch from aerobic to mixed aerobic/fermentation metabolism with increasing the

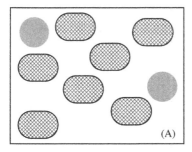 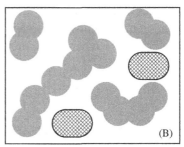

Figure 5.2 Impact of inert protein aggregates (grey circles) on the space available to allocate mitochondria (mesh ovals). (A) Cartoon of a neuron cell from healthy brain tissue. (B) Cartoon of a neuron cell from disease tissue of a patient with a neurodegenerative disease.

metabolic rate. In the case of running, the increased metabolic rate was associated with an increased energy demand for physical activity. In the case of cell growth the increased metabolic rate was associated with an increased energy demand for biosynthesis. In contrast, we can fairly assume that the energy demand in the brain is approximately constant. Or at least, we do not expect the brain energy demand to have such variations as observed from rest to fast running. Yet, we should bear in mind that the switch is also determined by the amount of volume that is available to allocate mitochondria. And this is where the formation of inert protein aggregates can have an impact on brain metabolism.

The formation of protein aggregates will reduce the space that is available for allocation of mitochondria. We could imagine the extreme case where almost all the brain tissue is occupied by protein aggregates (Fig. 5.2). In this hypothetical scenario, there is literally no room for aerobic metabolism and fermentation is required to satisfy the energy demand of brain activity. To provide a quantitative understanding of the impact of protein aggregates on brain metabolism, we will focus on the following model.

1. We will assume that the energy demand of brain metabolism is fixed to a value J_E, given in units of mol ATP/h/kg of body weight.
2. This energy demand can be satisfied by aerobic metabolism or fermentation metabolism

$$J_E = r_{AM} + r_F \tag{5.5}$$

where r_{AM} and r_F are the rates of aerobic and fermentation metabolism at the brain per unit of body weight.

3. The rates of aerobic and fermentation metabolism are determined by the corresponding horsepower and volume fractions of mitochondria and fermentation enzymes

$$r_{AM} = v_B \phi_M h_M \tag{5.6}$$

$$r_F = v_B \phi_F h_F \tag{5.7}$$

4. The blood lactate concentration associated with the fermentation component, as obtained from Eq. (3.15), is given by

$$C_L = \frac{W}{Q\beta_P} r_F \tag{5.8}$$

5. The overall volume fraction of mitochondria, fermentation enzymes and protein aggregates in the brain cannot exceed the maximum allowed tissue volume fraction ϕ_{max}

$$\phi_M + \phi_F + \phi_A \leq \phi_{max} \tag{5.9}$$

where ϕ_{max} is about 25% [57, 58].

As for the case of muscle metabolism, we will assume that the optimal solution is the combination of (r_{AM}, r_F) that minimises the lactate concentration in blood. For models (1)–(6), the optimal flux distribution (r_A, r_F) is given by

$$r_A = \begin{cases} J_E & 0 \leq \phi_A < \phi_N \\ J_E - \varepsilon_\phi(\phi_A - \phi_N) & \phi_N \leq \phi_A \leq \phi_{max} \end{cases} \tag{5.10}$$

$$r_F = \begin{cases} 0 & 0 \leq \phi_A < \phi_N \\ \varepsilon_\phi(\phi_A - \phi_N) & \phi_N \leq \phi_A \leq \phi_{max} \end{cases} \tag{5.11}$$

$$C_L = \begin{cases} 0 & 0 \leq \phi_A < \phi_N \\ \lambda_\phi(\phi_A - \phi_N) & \phi_N \leq \phi_A \leq \phi_{max} \end{cases} \tag{5.12}$$

$$\varepsilon_\phi = v_B h_M \tag{5.13}$$

$$\phi_N = \phi_{max} - \frac{j_E}{h_M} \tag{5.14}$$

$$j_E = \frac{J_E}{v_B} \tag{5.15}$$

$$\lambda_\phi = \frac{2W\varepsilon_\phi}{3Q\beta_P} \tag{5.16}$$

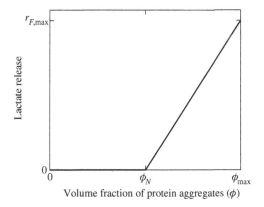

Figure 5.3 Lactate release as a function of the volume fraction of protein aggregates.

where j_E is the brain energy demand per unit of brain volume and ϕ_N is our prediction for the 'metabolic' onset of neurodegeneration.

This solution is graphically depicted in Fig. 5.3. When there is enough room for mitochondria in the brain tissue ($\phi_A < \phi_N$), the optimal solution is to satisfy the brain energy demand with aerobic metabolism. This holds until the protein aggregates occupy a volume fraction ϕ_N given by Eq. (5.14). Above this volume fraction of protein aggregates the brain cannot satisfy its energy demand with aerobic metabolism and fermentation must takes place. For protein aggregate volume fractions above ϕ_N, the rate of fermentation increases linearly with increasing the volume fraction of protein aggregates.

The prediction of a decrease of brain aerobic metabolism as the concentration of protein aggregates increases agrees with the experimental observation of a lower content of mitochondria in the neurons of Alzheimer's [127, 140] and Huntington's [128] patients. High lactate levels have been observed in certain brain regions of Huntington's disease patients compared to normal controls [135, 136], supporting the prediction of increased fermentation. However, these are just preliminary observations. Experimental studies should be conducted to determine whether neurodegeneration is actually associated with a switch from aerobic to mixed aerobic/fermentation metabolism.

To get an idea of the protein aggregates volume fraction that is required for the predicted metabolic switch let us estimate the threshold ϕ_N. Using reports of the brain energy demand [141] and the

Table 5.2 Parameter Estimation for Brain Metabolism, Including the Onset to Metabolic Neurodegeneration (ϕ_N)

Parameter	Value	Units	Reference
j_E	0.5	mol ATP/l/h	7.8 μmol/min/g [141], × 1 g/mL × 60 min/h
ϕ_{max}	25	% of tissue	[57, 58]
h_M	12	mol ATP/l/h	Median of healthy mammalian tissues, Table 2.1
j_E/h_M	4	% of tissue	
ϕ_N	21	% of tissue	Eq. (3.11)
h_M	3	mol ATP/l/h	Median of cancer cells, Table 2.1
j_E/h_M	17	% of tissue	
ϕ_N	8	% of tissue	Eq. (3.11)

mitochondria horsepower, we obtain $\phi_N \sim 21\%$ of tissue (Table 5.2). That is, when protein aggregates exceed 21% of the brain tissue, there will be a switch to mixed aerobic/fermentation metabolism. To estimate the threshold protein aggregation volume fraction that induces overflow metabolism in the brain (ϕ_N), we have assumed a typical mitochondria horsepower as deduced from healthy tissue. However, we cannot exclude that the mitochondria horsepower may be reduced in brain areas with protein aggregates. If the brain mitochondria of patients with neurodegenerative disorders has reduced horsepower relative to healthy tissue, which will result in a reduction of the protein aggregates volume fraction threshold as dictated by Eq. (5.14). For example, if we assume that mitochondria from disease brains has a horsepower as low as found in cancer cells (~ 3 mol ATP/l/h, Table 2.1), the protein aggregates volume fraction threshold goes down from 21% to 8% of brain tissue (Table 5.2).

Further, experimental studies are needed to test the validity of these predictions. That should include measurements of the mitochondria horsepower from disease brain tissue, the cell volume fraction occupied by inert protein aggregates and the rate of lactate release from the disease brain tissue.

CHAPTER 6

Outlook

The calculations presented here demonstrate, we can get a good estimate of maximum metabolic rates using as input the specific biochemical horsepower and basic geometric considerations. The specific biochemical horsepower – mol of metabolic product/litre of metabolic machinery/hour – measures the intrinsic metabolic output of the metabolic machinery. The geometric considerations are required to determine how tight the metabolic machinery can be packed. In turn, because different metabolic pathways have different specific horsepower, there will be transitions from one to the next in increasing order of their horsepower as the metabolic demand is increased.

In the context of energy metabolism, this is exemplified by oxidative phosphorylation and fermentation. Oxidative phosphorylation has a lower specific horsepower for energy production than fermentation and consequently fermentation can satisfy higher energy demands than oxidative phosphorylation. Therefore, under nutrient-rich conditions or high energy demands fermentation will be selected for. The micro or multicellular organism will benefit from the evolution of regulatory mechanisms that would switch on fermentation when the rich nutrient conditions or high energy demands are met. That explains the widespread occurrence of overflow metabolism.

It is compelling to extrapolate that similar metabolic switches could be taking place in other pathways of cell and whole body metabolism. Going down the list of major metabolic requirements, we could anticipate metabolic switches in the context of nitrogen, lipid, one-carbon and nucleotide metabolism. All these metabolic pathways contain more than one alternative to provide the end product. And most certainly, the different alternatives will have different yields and horsepowers. Any instance where one metabolic route has higher yield and lower horsepower than another route would be a candidate for a metabolic switch. Furthermore, since lower metabolic yield generally takes

Overflow Metabolism. DOI: https://doi.org/10.1016/B978-0-12-812208-2.00006-8

place concomitant with metabolic byproduct release, those metabolic switches would be candidates for overflow metabolism.

The analysis presented here about aerobic fermentation in human disease is quite preliminary. There is much we can learn by linking the metabolic output of the disease tissue or immune response to the burden caused by that output on the whole body metabolism. Many of the diseases considered are or can become chronic diseases and therefore such metabolic burden would be sustained for long period of times. Examples where this could apply include autoimmune diseases, chronic inflammation, cancer and neurodegenerative diseases.

REFERENCES

[1] Voet D, Voet JG. 4th ed. Biochemistry, xxv. Hoboken, NJ: John Wiley & Sons, Inc; 2011. 3, 1428, 53 pp.

[2] Porter JR. Louis PASTEUR; achievements and disappointments, 1861. Bacteriol Rev 1961;25:389−403.

[3] Pasteur ML. Etudes sur la biere. Paris: Gauthier-Villars; 1876.

[4] Fletcher WM. Lactic acid in amphibian muscle. J Physiol 1907;35(4):247−309.

[5] Brown AJ. Influence of oxygen and concentration on alcoholic fermentation. J Chem Soc Trans 1892;61:369−85.

[6] Brown AJ. The specific character of the fermentative functions of yeast cells. J Chem Soc Trans 1894;65:911−23.

[7] Briant L. The influence of aëration on fermentation. J Federat Inst Brewing 1895;1 (5):472−86.

[8] Warburg O, Minami S. Versuche an Überlebendem Carcinomgewebe. Klin Wochenschr 1923;2(17):776−7.

[9] Crabtree HG. The carbohydrate metabolism of certain pathological overgrowths. Biochem J 1928;22(5):1289−98.

[10] Warburg O. Über den heutigen Stand des Carcinomproblems. Naturwissenschaften 1927;15 (1):1−4.

[11] Warburg O. On the origin of cancer cells. Science 1956;123(3191):309−14.

[12] Crabtree HG. Observations on the carbohydrate metabolism of tumours. Biochem J 1929;23 (3):536−45.

[13] Anand SK, Tikoo SK. Viruses as modulators of mitochondrial functions. Adv Virol 2013;2013:738−94.

[14] Swanson WH, Clifton CE. Growth and assimilation in cultures of *Saccharomyces cerevisiae*. J Bacteriol 1948;56(1):115−24.

[15] De Deken RH. The Crabtree effect: a regulatory system in yeast. J General Microbiol 1966;44(2):149−56.

[16] Owles WH. Alterations in the lactic acid content of the blood as a result of light exercise, and associated changes in the co(2)-combining power of the blood and in the alveolar co(2) pressure. J Physiol 1930;69(2):214−37.

[17] Needham DM. Machina carnis: the biochemistry of muscular contraction and its historical development. London: Cambridge University Press; 1971. xvi, 782, 20 pp.

[18] Wasserman K, McIlroy MB. Detecting the threshold of anaerobic metabolism in cardiac patients during exercise. Am J Cardiol 1964;14:844−52.

[19] Foster JW. Some introspections on mold metabolism. Bacteriol Rev 1947;11(3):167−88.

[20] Andersen KB, von Meyenburg K. Are growth rates of *Escherichia coli* in batch cultures limited by respiration? J Bacteriol 1980;144(1):114−23.

[21] Majewski RA, Domach MM. Simple constrained-optimization view of acetate overflow in *E. coli*. Biotechnol Bioeng 1990;35(7):732–8.

[22] Varma A, Palsson BO. Stoichiometric flux balance models quantitatively predict growth and metabolic by-product secretion in wild-type *Escherichia coli* W3110. Appl Environ Microbiol 1994;60(10):3724–31.

[23] Bucher M, Brander KA, Sbicego S, Mandel T, Kuhlemeier C. Aerobic fermentation in tobacco pollen. Plant Molec Biol 1995;28(4):739–50.

[24] Tadege M, Kuhlemeier C. Aerobic fermentation during tobacco pollen development. Plant Molec Biol 1997;35(3):343–54.

[25] el-Mansi EM, Holms WH. Control of carbon flux to acetate excretion during growth of *Escherichia coli* in batch and continuous cultures. J General Microbiol 1989;135 (11):2875–83.

[26] Stouthamer AH, Bettenhaussen CW. Determination of the efficiency of oxidative phosphorylation in continuous cultures of *Aerobacter aerogenes*. Arch Microbiol 1975;102(3):187–92.

[27] Neijssel OM, Tempest DW. The regulation of carbohydrate metabolism in *Klebsiella aerogenes* NCTC 418 organisms, growing in chemostat culture. Arch Microbiol 1975;106 (3):251–8.

[28] Russell JB, Baldwin RL. Comparison of maintenance energy expenditures and growth yields among several rumen bacteria grown on continuous culture. Appl Environ Microbiol 1979;37(3):537–43.

[29] Cazzulo JJ, Decazzulo BMF, Engel JC, Cannata JJB. End products and enzyme levels of aerobic glucose fermentation in trypanosomatids. Mol Biochem Parasit 1985;16(3):329–43.

[30] de Bari L, Moro L, Passarella S. Prostate cancer cells metabolize d-lactate inside mitochondria via a D-lactate dehydrogenase which is more active and highly expressed than in normal cells. Febs Lett 2013;587(5):467–73.

[31] Kioka H, Kato H, Fujikawa M, Tsukamoto O, Suzuki T, Imamura H, et al. Evaluation of intramitochondrial ATP levels identifies G0/G1 switch gene 2 as a positive regulator of oxidative phosphorylation. Proc Natl Acad Sci USA 2014;111(1):273–8.

[32] Short KR, Nygren J, Barazzoni R, Levine J, Nair KS. T(3) increases mitochondrial ATP production in oxidative muscle despite increased expression of UCP2 and -3. Am J Physiol Endocrinol Metab 2001;280(5):E761–9.

[33] Karakelides H, Irving BA, Short KR, O'Brien P, Nair KS. Age, obesity, and sex effects on insulin sensitivity and skeletal muscle mitochondrial function. Diabetes 2010;59(1):89–97.

[34] Yoshioka J, Chutkow WA, Lee S, Kim JB, Yan J, Tian R, et al. Deletion of thioredoxin-interacting protein in mice impairs mitochondrial function but protects the myocardium from ischemia-reperfusion injury. J Clin Invest 2012;122(1):267–79.

[35] Hou XY, Green S, Askew CD, Barker G, Green A, Walker PJ. Skeletal muscle mitochondrial ATP production rate and walking performance in peripheral arterial disease. Clin Physiol Funct Imaging 2002;22(3):226–32.

[36] Gonzalvez F, Pariselli F, Dupaigne P, Budihardjo I, Lutter M, Antonsson B, et al. tBid interaction with cardiolipin primarily orchestrates mitochondrial dysfunctions and subsequently activates Bax and Bak. Cell Death Differ 2005;12(6):614–26.

[37] Vazquez A, Liu J, Zhou Y, Oltvai ZN. Catabolic efficiency of aerobic glycolysis: the Warburg effect revisited. Bmc Syst Biol 2010;4:58.

[38] Vazquez A. Limits of aerobic metabolism in cancer cells. 2013. http://dx.doi.org/10.1101/020461.

[39] Aste T, Weaire DL. 2nd ed. The pursuit of perfect packing, xiii. New York, London: Taylor & Francis; 2008. 200 pp.

[40] Hales TC. A proof of the Kepler conjecture. Ann Math 2005;162(3):1065–185.

[41] Scott GD, Kilgour DM. Density of random close packing of spheres. J Phys D Appl Phys 1969;2(6):863–6.

[42] Gotoh K, Finney JL. Statistical geometrical approach to random packing density of equal spheres. Nature 1974;252(5480):202–5.

[43] Sherwood JD. Packing of spheroids in three-dimensional space by random sequential addition. J Phys a-Math Gen 1997;30(24):L839–43.

[44] Zangmeister CD, Radney JG, Dockery LT, Young JT, Ma X, You R, et al. Packing density of rigid aggregates is independent of scale. Proc Natl Acad Sci USA 2014;111(25):9037–41.

[45] Posakony JW, England JM, Attardi G. Mitochondrial growth and division during the cell cycle in HeLa cells. J Cell Biol 1977;74(2):468–91.

[46] Singh I, Tsang KY, Ludwig GD. Alterations in the mitochondria of human osteosarcoma cells with glucocorticoids. Cancer Res 1974;34(11):2946–52.

[47] DiSorbo DM, Paavola LG, Litwack G. Pyridoxine resistance in a rat hepatoma cell line. Cancer Res 1982;42(6):2362–70.

[48] Sesso A, Marques MM, Monteiro MMT, Schumacher RI, Colquhoun A, Belizario J, et al. Morphology of mitochondrial permeability transition: morphometric volumetry in apoptotic cells. Anat Rec Part A 2004;281A(2):1337–51.

[49] Bertoni-Freddari C, Fattoretti P, Giorgetti B, Grossi Y, Balietti M, Casoli T, et al. Synaptic and mitochondrial morphometry provides structural correlates of successful brain aging. Ann NY Acad Sci 2007;1097:51–3.

[50] Yuan H, Gerencser AA, Liot G, Lipton SA, Ellisman M, Perkins GA, et al. Mitochondrial fission is an upstream and required event for Bax foci formation in response to nitric oxide in cortical neurons. Cell Death Differ 2007;14(3):462–71.

[51] Lauschova I, Krejcirova L, Horky D, Doubek M, Mayer J, Doubek J. Ultrastructural morphometry of renal tubule epithelium in rats treated with conventional amphotericin B deoxycholate or amphotericin B colloidal dispersion. Acta Vet Brno 2004;73(2):165–9.

[52] Sullivan SM, Pittman RN. Relationship between mitochondrial volume density and capillarity in hamster muscles. Am J Physiol 1987;252(1):H149–55.

[53] Barth E, Stammler G, Speiser B, Schaper J. Ultrastructural quantitation of mitochondria and myofilaments in cardiac muscle from 10 different animal species including man. J Molec Cell Cardiol 1992;24(7):669–81.

[54] Urschel MR, O'Brien KM. High mitochondrial densities in the hearts of Antarctic icefishes are maintained by an increase in mitochondrial size rather than mitochondrial biogenesis. J Exp Biol 2008;211(Pt 16):2638–46.

[55] O'Brien KM, Sidell BD. The interplay among cardiac ultrastructure, metabolism and the expression of oxygen-binding proteins in Antarctic fishes. J Exp Biol 2000;203 (Pt 8):1287–97.

[56] Suarez RK, Lighton JR, Brown GS, Mathieu-Costello O. Mitochondrial respiration in hummingbird flight muscles. Proc Natl Acad Sci USA 1991;88(11):4870–3.

[57] Robertson JD. Studies on the chemical composition of muscle tissue. 3. The mantle muscle of cephalopod molluscs. J Exp Biol 1965;42:153–75.

[58] Bergstrom J, Furst P, Noree LO, Vinnars E. Intracellular free amino acid concentration in human muscle tissue. J Appl Physiol 1974;36(6):693–7.

[59] Kim YR, Savellano MD, Savellano DH, Weissleder R, Bogdanov A. Measurement of tumor interstitial volume fraction: Method and implication for drug delivery. Magnet Reson Med 2004;52(3):485–94.

[60] Wilmore JH, Costill DL, Kenney WL. 6th ed Physiology of sport and exercise, xix. Champaign, IL: Human Kinetics; 2015. 624 pp.

[61] Taylor CR, Schmidt-Nielsen K, Raab JL. Scaling of energetic cost of running to body size in mammals. Am J Phys 1970;219(4):1104–7.

[62] Mcmahon TA. Mechanics of locomotion. Int J Robot Res 1984;3(2):4–28.

[63] Janssen I, Heymsfield SB, Wang ZM, Ross R. Skeletal muscle mass and distribution in 468 men and women aged 18-88 yr. J Appl Physiol 2000;89(1):81–8.

[64] Wishart DS, Jewison T, Guo AC, Wilson M, Knox C, Liu Y, et al. HMDB 3.0—the human metabolome database in 2013. Nucleic Acids Res 2013;41(Database issue):D801–7.

[65] Farrell PA, Wilmore JH, Coyle EF, Billing JE, Costill DL. Plasma lactate accumulation and distance running performance. Med Sci Sports 1979;11(4):338–44.

[66] Neidhardt FC, Ingraham JL, Schaechter M. Physiology of the bacterial cell: a molecular approach. Sunderland, MA: Sinauer Associates; 1990.

[67] Forster J, Famili I, Fu P, Palsson BO, Nielsen J. Genome-scale reconstruction of the *Saccharomyces cerevisiae* metabolic network. Genome Res 2003;13(2):244–53.

[68] Alberts B. Molecular biology of the cell. 5th ed. New York: Garland Science; 2008. 1 v. (various pagings) p.

[69] Bremer H, Dennis PP. Modulation of chemical composition and other parameters of the cell at different exponential growth rates. EcoSal Plus 2008. Available from: http://dx.doi.org/10.1128/ecosal.5.2.3.

[70] Gabashvili IS, Agrawal RK, Spahn CM, Grassucci RA, Svergun DI, Frank J, et al. Solution structure of the E. coli 70S ribosome at 11.5 A resolution. Cell 2000;100(5):537–49.

[71] Zimmerman SB, Trach SO. Estimation of macromolecule concentrations and excluded volume effects for the cytoplasm of *Escherichia coli*. J Mol Biol 1991;222(3):599–620.

[72] Basan M, Zhu M, Dai X, Warren M, Sevin D, Wang YP, et al. Inflating bacterial cells by increased protein synthesis. Mol Syst Biol 2015;11(10):836.

[73] Record Jr. MT, Courtenay ES, Cayley DS, Guttman HJ. Responses of E. coli to osmotic stress: large changes in amounts of cytoplasmic solutes and water. Trends Biochem Sci 1998;23(4):143–8.

[74] Wittmann HG. Components of bacterial ribosomes. Annu Rev Biochem 1982;51:155–83.

[75] Schmidt A, Kochanowski K, Vedelaar S, Ahrne E, Volkmer B, Callipo L, et al. The quantitative and condition-dependent *Escherichia coli* proteome. Nat Biotechnol 2016;34(1):104–10.

[76] Brocchieri L, Karlin S. Protein length in eukaryotic and prokaryotic proteomes. Nucleic Acids Res 2005;33(10):3390–400.

[77] Savinell JM, Palsson BO. Network analysis of intermediary metabolism using linear optimization. I. Development of mathematical formalism. J Theor Biol 1992;154(4):421–54.

[78] Ingolia NT, Lareau LF, Weissman JS. Ribosome profiling of mouse embryonic stem cells reveals the complexity and dynamics of mammalian proteomes. Cell 2011;147(4):789–802.

[79] Khatter H, Myasnikov AG, Natchiar SK, Klaholz BP. Structure of the human 80S ribosome. Nature 2015;520(7549):640–5.

[80] Menetret JF, Neuhof A, Morgan DG, Plath K, Radermacher M, Rapoport TA, et al. The structure of ribosome-channel complexes engaged in protein translocation. Mol Cell 2000;6(5):1219–32.

[81] Sheikh K, Forster J, Nielsen LK. Modeling hybridoma cell metabolism using a generic genome-scale metabolic model of *Mus musculus*. Biotechnol Prog 2005;21(1):112–21.

[82] Dolfi SC, Chan LL, Qiu J, Tedeschi PM, Bertino JR, Hirshfield KM, et al. The metabolic demands of cancer cells are coupled to their size and protein synthesis rates. Cancer Metabol 2013;1(1):20.

[83] Kilburn DG, Lilly MD, Webb FC. The energetics of mammalian cell growth. J Cell Sci 1969;4(3):645−54.

[84] De Boer RJ, Homann D, Perelson AS. Different dynamics of CD4+ and CD8+ T cell responses during and after acute lymphocytic choriomeningitis virus infection. J Immunol 2003;171(8):3928−35.

[85] Wang T, Marquardt C, Foker J. Aerobic glycolysis during lymphocyte proliferation. Nature 1976;261(5562):702−5.

[86] Hume DA, Radik JL, Ferber E, Weidemann MJ. Aerobic glycolysis and lymphocyte transformation. Biochem J 1978;174(3):703−9.

[87] Nilsson A, Nielsen J. Metabolic trade-offs in yeast are caused by F1F0-ATP synthase. Sci Rep 2016;6:22264.

[88] Lecault V, Patel N, Thibault J. An image analysis technique to estimate the cell density and biomass concentration of Trichoderma reesei. Lett Appl Microbiol 2009;48(4):402−7.

[89] Smith HL, Waltman P. The theory of the chemostat: dynamics of microbial competition. New York: Cambridge University Press; 1995.

[90] Van Hoek P, Van Dijken JP, Pronk JT. Effect of specific growth rate on fermentative capacity of baker's yeast. Appl Environ Microbiol 1998;64(11):4226−33.

[91] Boulton RB, Singleton VL, Bisson LF, Kunkee RE. Principles and practices of winemaking. New York: Springer; 1999.

[92] Basan M, Hui S, Okano H, Zhang Z, Shen Y, Williamson JR, et al. Overflow metabolism in Escherichia coli results from efficient proteome allocation. Nature 2015;528 (7580):99−104.

[93] Vazquez A, Oltvai ZN. Macromolecular crowding explains overflow metabolism in cells. Sci Rep 2016;6:31007.

[94] Feist AM, Herrgard MJ, Thiele I, Reed JL, Palsson BO. Reconstruction of biochemical networks in microorganisms. Nat Rev Microbiol 2009;7(2):129−43.

[95] Price ND, Reed JL, Palsson BO. Genome-scale models of microbial cells: evaluating the consequences of constraints. Nat Rev Microbiol 2004;2(11):886−97.

[96] Lewis NE, Nagarajan H, Palsson BO. Constraining the metabolic genotype-phenotype relationship using a phylogeny of in silico methods. Nat Rev Microbiol 2012;10(4):291−305.

[97] Lewis NE, Abdel-Haleem AM. The evolution of genome-scale models of cancer metabolism. Front Physiol 2013;4:237.

[98] Beg QK, Vazquez A, Ernst J, de Menezes MA, Bar-Joseph Z, Barabasi AL, et al. Intracellular crowding defines the mode and sequence of substrate uptake by Escherichia coli and constrains its metabolic activity. Proc Natl Acad Sci USA 2007;104(31):12663−8.

[99] Adadi R, Volkmer B, Milo R, Heinemann M, Shlomi T. Prediction of microbial growth rate versus biomass yield by a metabolic network with kinetic parameters. PLoS Comput Biol 2012;8(7):e1002575.

[100] Vazquez A, Beg QK, Demenezes MA, Ernst J, Bar-Joseph Z, Barabasi AL, et al. Impact of the solvent capacity constraint on E. coli metabolism. Bmc Syst Biol 2008;2:7.

[101] Molenaar D, van Berlo R, de Ridder D, Teusink B. Shifts in growth strategies reflect trade-offs in cellular economics. Mol Syst Biol 2009;5:323.

[102] Shlomi T, Benyamini T, Gottlieb E, Sharan R, Ruppin E. Genome-scale metabolic modeling elucidates the role of proliferative adaptation in causing the warburg effect. PLoS Comput Biol 2011;7(3):e1002018.

[103] Vazquez A, Markert EK, Oltvai ZN. Serine biosynthesis with one carbon catabolism and the glycine cleavage system represents a novel pathway for ATP generation. PLos One 2011;6(11):e25881.

[104] Capuani F, De Martino D, Marinari E, De Martino A. Quantitative constraint-based computational model of tumor-to-stroma coupling via lactate shuttle. Sci Rep 2015;5:11880.

[105] Fernandez-de-Cossio-Diaz J, De Martino A, Mulet R. Microenvironmental cooperation promotes early spread and bistability of a Warburg-like phenotype. Sci Rep 2017;7(1):3103.

[106] Martin-Jimenez CA, Salazar-Barreto D, Barreto GE, Gonzalez J. Genome-scale reconstruction of the human astrocyte metabolic network. Front Aging Neurosci 2017;9:23.

[107] Yizhak K, Benyamini T, Liebermeister W, Ruppin E, Shlomi T. Integrating quantitative proteomics and metabolomics with a genome-scale metabolic network model. Bioinformatics 2010;26(12):i255−60.

[108] Villaverde AF, Egea JA, Banga JR. A cooperative strategy for parameter estimation in large scale systems biology models. Bmc Syst Biol 2012;6.

[109] Smallbone K, Simeonidis E, Swainston N, Mendes P. Towards a genome-scale kinetic model of cellular metabolism. Bmc Syst Biol 2010;4:6.

[110] Stanford NJ, Lubitz T, Smallbone K, Klipp E, Mendes P, Liebermeister W. Systematic construction of kinetic models from genome-scale metabolic networks. PLoS One 2013;8 (11):e79195.

[111] Muller S, Regensburger G, Steuer R. Enzyme allocation problems in kinetic metabolic networks: optimal solutions are elementary flux modes. J Theor Biol 2014;347:182−90.

[112] Wortel MT, Peters H, Hulshof J, Teusink B, Bruggeman FJ. Metabolic states with maximal specific rate carry flux through an elementary flux mode. FEBS J 2014;281 (6):1547−55.

[113] Schuster S, Fell DA, Dandekar T. A general definition of metabolic pathways useful for systematic organization and analysis of complex metabolic networks. Nat Biotechnol 2000;18(3):326−32.

[114] Bakker J, Gris P, Coffernils M, Kahn RJ, Vincent JL. Serial blood lactate levels can predict the development of multiple organ failure following septic shock. Am J Surg 1996;171 (2):221−6.

[115] Fischer K, Hoffmann P, Voelkl S, Meidenbauer N, Ammer J, Edinger M, et al. Inhibitory effect of tumor cell-derived lactic acid on human T cells. Blood 2007;109(9):3812−19.

[116] Bakker J, Nijsten MW, Jansen TC. Clinical use of lactate monitoring in critically ill patients. Ann Intens Care 2013;3(1):12.

[117] Tisdale MJ. Mechanisms of cancer cachexia. Physiol Rev 2009;89(2):381−410.

[118] Trojanowski JQ, Goedert M, Iwatsubo T, Lee VM. Fatal attractions: abnormal protein aggregation and neuron death in Parkinson's disease and Lewy body dementia. Cell Death Different 1998;5(10):832−7.

[119] Ross CA, Poirier MA. Protein aggregation and neurodegenerative disease. Nat Med 2004;10:Suppl:S10−17.

[120] Irvine GB, El-Agnaf OM, Shankar GM, Walsh DM. Protein aggregation in the brain: the molecular basis for Alzheimer's and Parkinson's diseases. Mol Med 2008;14(7-8):451−64.

[121] Wanker EE. Protein aggregation and pathogenesis of Huntington's disease: mechanisms and correlations. Biol Chem 2000;381(9—10):937—42.

[122] Hoffner G, Djian P. Protein aggregation in Huntington's disease. Biochimie 2002;84 (4):273—8.

[123] Askanas V, Engel WK. Inclusion-body myositis and myopathies: different etiologies, possibly similar pathogenic mechanisms. Curr Opin Neurol 2002;15(5):525—31.

[124] Roos PM, Vesterberg O, Nordberg M. Inclusion body myositis in Alzheimer's disease. Acta Neurol Scand 2011;124(3):215—17.

[125] Roth J, Yam GH, Fan J, Hirano K, Gaplovska-Kysela K, Le Fourn V, et al. Protein quality control: the who's who, the where's and therapeutic escapes. Histochem Cell Biol 2008;129(2):163—77.

[126] Ceru S, Layfield R, Zavasnik-Bergant T, Repnik U, Kopitar-Jerala N, Turk V, et al. Intracellular aggregation of human stefin B: confocal and electron microscopy study. Biol Cell 2010;102(6):319—34.

[127] Baloyannis SJ, Costa V, Michmizos D. Mitochondrial alterations in Alzheimer's disease. Am J Alzheimers Dis Other Demen 2004;19(2):89—93.

[128] Kim J, Moody JP, Edgerly CK, Bordiuk OL, Cormier K, Smith K, et al. Mitochondrial loss, dysfunction and altered dynamics in Huntington's disease. Hum Mol Genet 2010;19 (20):3919—35.

[129] Sheng B, Wang X, Su B, Lee HG, Casadesus G, Perry G, et al. Impaired mitochondrial biogenesis contributes to mitochondrial dysfunction in Alzheimer's disease. J Neurochem 2012;120(3):419—29.

[130] Trushina E, Nemutlu E, Zhang S, Christensen T, Camp J, Mesa J, et al. Defects in mitochondrial dynamics and metabolomic signatures of evolving energetic stress in mouse models of familial Alzheimer's disease. PLoS One 2012;7(2):e32737.

[131] Kopeikina KJ, Carlson GA, Pitstick R, Ludvigson AE, Peters A, Luebke JI, et al. Tau accumulation causes mitochondrial distribution deficits in neurons in a mouse model of tauopathy and in human Alzheimer's disease brain. Am J Pathol 2011;179(4):2071—82.

[132] Hu Y, Wilson GS. A temporary local energy pool coupled to neuronal activity: fluctuations of extracellular lactate levels in rat brain monitored with rapid-response enzyme-based sensor. J Neurochem 1997;69(4):1484—90.

[133] Mangia S, Garreffa G, Bianciardi M, Giove F, Di Salle F, Maraviglia B. The aerobic brain: lactate decrease at the onset of neural activity. Neuroscience 2003;118(1):7—10.

[134] Kasischke KA, Vishwasrao HD, Fisher PJ, Zipfel WR, Webb WW. Neural activity triggers neuronal oxidative metabolism followed by astrocytic glycolysis. Science 2004;305 (5680):99—103.

[135] Jenkins BG, Koroshetz WJ, Beal MF, Rosen BR. Evidence for impairment of energy metabolism in vivo in Huntington's disease using localized 1H NMR spectroscopy. Neurology 1993;43(12):2689—95.

[136] Harms L, Meierkord H, Timm G, Pfeiffer L, Ludolph AC. Decreased N-acetyl-aspartate/choline ratio and increased lactate in the frontal lobe of patients with Huntington's disease: a proton magnetic resonance spectroscopy study. J Neurol Neurosurg Psychiatry 1997;62 (1):27—30.

[137] Vazquez A. Metabolic states following accumulation of intracellular aggregates: implications for neurodegenerative diseases. PLoS One 2013;8(5):e63822.

[138] Minton AP. The influence of macromolecular crowding and macromolecular confinement on biochemical reactions in physiological media. J Biol Chem 2001;276(14):10577—80.

[139] Vazquez A. Optimal cytoplasmatic density and flux balance model under macromolecular crowding effects. J Theor Biol 2010;264(2):356−9.

[140] Hirai K, Aliev G, Nunomura A, Fujioka H, Russell RL, Atwood CS, et al. Mitochondrial abnormalities in Alzheimer's disease. J Neurosci 2001;21(9):3017−23.

[141] Chaumeil MM, Valette J, Guillermier M, Brouillet E, Boumezbeur F, Herard AS, et al. Multimodal neuroimaging provides a highly consistent picture of energy metabolism, validating 31P MRS for measuring brain ATP synthesis. Proc Natl Acad Sci USA 2009;106 (10):3988−93.

Printed in the United States
By Bookmasters